HUMAN AND ANIMAL-POWERED WATER-LIFTING DEVICES

W. K. Kennedy and T. A. Rogers

Intermediate Technology Publications 1985

Intermediate Technology Publications 1985
9 King Street, London WC2E 8HW, U.K.

© Intermediate Technology Publications 1985
ISBN 0 946688 75 3

Printed in England by Print Power (London) Limited
7 Willow Street, London EC2A 4QH, UK

PREFACE

This survey is the outcome of a desk study aimed at determining possibilities for an ITDG project in the field of water-lifting by human and animal-power. Considering the attention which has been given to other renewable energy pumping systems - wind-, solar-, water- and biomass-powered - it was felt that human and animal power were neglected subject areas which deserved greater study. Our findings have confirmed that the quantity, quality and reliability of the data available leave a lot to be desired and one of the purposes of publishing this survey is to encourage individuals to help us correct these deficiencies by providing us with additional accurate and detailed information. A reply address is given at the end of the survey for this purpose and we hope that you, the readers, will share your experience with us so that others can benefit.

From the outset of the study - which was carried out by Bill Kennedy (on sabbatical from the Open University) with help from Traude Rogers - we were, of course, aware of the large amount of research and development being done on handpumps for rural drinking water supply. Our brief was, of course, much wider than this and we were particularly interested in the use of water for productive purposes, where there might be more potential for raising local incomes and improving livelihoods. The water-lifting device *per se* was recognized as being only one of the components of the whole water delivery system and so we have tried to avoid looking exclusively at that part.

The authors have attempted to produce a balanced view of the subject from both the technical and socio-economic perspectives and we hope that the survey will interest rural development workers involved in water supply for domestic and for agricultural purposes. The technical chapters of the survey are illustrated, so that the reader can gain an impression of how different devices are made and operated. Typical performance data is provided where available, and graphs or tables have been produced to assist comparison.

It is hoped that this report will stimulate feedback to ITDG, feedback which in the longer term may help to define more clearly the priorities for field project activities. Finally, ITDG is extremely grateful to Bill Kennedy and Traude Rogers for their hard work and patience, and thanks are due to Rosalyn Burne who helped to finalize the text.

John Collett,
Water Officer, ITDG, Rugby, U.K.

CONTENTS

		Page
PART 1	INTRODUCTION	1
PART 2	HUMAN-POWERED WATER-LIFTERS: TECHNICAL ASPECTS	2
2.1	Human power: what power is available?	2
2.2	Review of traditional water-lifting devices	6
2.3	Practicable irrigated area of traditional water-lifters	22
2.4	Research and development on traditional water-lifting devices	22
2.5	Hand pump types	30
2.6	Hand pump types: assessment/relative merits	41
2.7	Hand pump testing	44
2.7.1	Hand pump laboratory testing carried out by the Consumers' Association	44
2.7.2	Results and conclusions of the Consumers' Association tests	49
2.7.3	Ergonomic aspects of pumps	52
2.8	Hand pump research and development	54
2.8.1	Various prototype pumps	54
2.8.2	A possible R & D project on human-powered irrigation pumps	70
PART 3	HUMAN-POWERED WATER-LIFTERS: SOCIAL ASPECTS	71
3.1	Introduction	71
3.2	Community water supply systems	71
3.3	Water supply for irrigation	77
PART 4	ANIMAL-POWERED WATER-LIFTERS: TECHNICAL ASPECTS	80
4.1	Animal power: what power is available?	80
4.2	Review of traditional animal-powered water-lifting devices	81
4.2.1	Rope and pulley arrangements	81
4.2.2	The circular mohte	84
4.2.3	The Persian wheel	84
4.2.4	Developments based on the Persian wheel	87
4.3	Research and development on traditional animal-powered devices	89
4.3.1	The animal-powered irrigation pump	89
4.3.2	Research and development at the Appropriate Technology Development Association, Lucknow, India	90

PART 5	DRAUGHT ANIMAL POWER FOR PUMPING: SOCIO-ECONOMIC ASPECTS	
5.1	Introduction	91
5.2	Draught animal power in Africa	92
5.3	Draught animal power in Asia	95
PART 6	IMPACT ASSESSMENT, DIFFUSION OF TECHNOLOGIES AND MAINTENANCE AND TRAINING PROGRAMMES	99
6.1	Introduction	99
6.2	Guidelines for project development	100
6.3	Maintenance and training programmes	101
6.4	Impact assessment - The case of the MOSTI	104
APPENDIX 1	Contact address	107
APPENDIX 2	References	108

1. INTRODUCTION

The purpose of this study is to provide a state-of-the-art survey of both the technical and the socio-economic aspects of human and animal-powered water-lifting. This report is intended to act as an 'ideas generator'; it was originally hoped that it would serve as an interim step to the formulation of research and development projects in human and animal-powered pumping. The report does indicate possible areas for development but, in addition, its style of presentation is such that at intervals it seeks to stimulate a response from the reader - whether it be by seeking further field data on a particular water-lifter, by asking for comment on a proposed development or by requesting possible project ideas. It is therefore hoped that the report will be a working document and will stimulate comment and discussion.

Why limit the study to human and animal-powered devices?

It was considered that in the area of rural energy supplies and associated power devices for water-lifting, human and animal power is often not covered adequately in the literature. Although developments in human- and animal-powered devices are often mooted there is little real progress in this area. Indeed much of the 'alternative energy' literature relates to energy sources at an 'intermediate' level between the traditional human/animal power and the 'conventional' (from a developed society viewpoint) usually 'high tech' utilization of energy sources. Hence energy for rural development is often associated with solar, wind, hydro, photosynthesis and bio-gas sources. Although developments in such renewable resources must be given due emphasis as alternatives to fossil fuels, it is nevertheless suggested that in rural areas there is often no real widespread alternative to human- and animal-powered water-lifters. The problem here is that much of the 'intermediate level' alternative technology is still not appropriate to many of the water-lifting problems of rural areas. Although technical solutions are proposed, the socio-economic factors often mitigate against any widespread adoption. Hence we have a dilemma:
- Is it possible that the intermediate technology solution for rural energy supplies is not appropriate simply because the level of the proposed technology is still too high and does not give due emphasis to the social and economic realities of rural areas?
- Is there scope for upgrading traditional water-lifting devices?

Within the initial thinking which motivated the present study was the suggestion that perhaps there might be a gap in the technology spectrum of water-lifters. Could there be scope for upgrading traditional devices which are often very inefficient in their use of human and animal power? Perhaps the technology 'gap' is one between traditional devices and many of the proposed 'intermediate' solutions? Without seeking to limit the power sources of rural areas to human and animal sources alone, it should be of value to assess fully possible developments in this area.

The importance of socio-economic aspects in any proposed developments in traditional water-lifters

If developments of traditional water-lifters are to be meaningful it will be essential to ensure that any upgraded devices would be appropriate to the socio-economic realities of the rural situation. It will be important not to 'up-market' the existing devices so much that people cannot afford them (even with credit facilities available), or that it may not be possible to replicate them widely. A sensitive re-appraisal of the existing technology coupled with a study of the social infrastructure would seem to be essential, and community participation would be essential at the pre-project stage in order to assess problems fully and identify fruitful areas for development. To some extent this report emphasizes socio-economic aspects of work in this area since it is accepted that no amount of technical (hardware) development is worthwhile which does not take socio-economic factors fully into account. Too much mere 'lip-service' is paid to this last point. As will be noted in this report it is positively recommended that no practical 'hardware' work be undertaken until a field study is undertaken which would include an extensive socio-economic assessment and a sensitive re-appraisal of existing methods of lifting water.

2. HUMAN-POWERED WATER-LIFTERS: TECHNICAL ASPECTS

2.1 Human power: what power is available?

In the rural areas of many developing countries the major water-lifting (and water-carrying) tasks are done by human effort - mainly that of women. The human power available is subject to considerable variation. It depends on:
- the stature/weight, age, sex, health and diet of the individual;
- the environmental conditions, temperature/humidity, altitude;
- ergonomic considerations, i.e. body posture during work (height of pump handle, length of stroke, etc);
- the muscle groups used in the activity, e.g. mainly arm/shoulders, more 'all body' use (arms, shoulders,

back, legs), pedalling;
- whether short stints (a few minutes, 10 to 15 minutes) or fairly sustained efforts over a working day (with rest periods).

With such variables it is not surprising that the literature gives conflicting values for human power capability (in many cases figures are quoted without specifying the relevant operating conditions). In a European climate a power range of 45 to 60W may be considered to be a reasonable 'all body' (i.e. not limited to arms/shoulders but also back, and perhaps legs, but not pedalling) sustained power output over a 5-hour period (with some rest periods). For tropical climates and with poor health/malnourishment of the populace this 'reasonable' range could well be halved. For the purposes of this study it is proposed to consider a reasonable power output of adults in favourable health in a tropical climate as follows:

Table 1. Typical adult power output

Muscle group(s) involved	Sustained (up to 6/7 hrs) per day with short rests as and when req'd	10 to 15 mins	Few mins
Mainly arms/shoulders	30W	50W	70W
All body (arms/shoulders, back, legs - non-pedalling)	40 to 50W	70W	100W
Pedalling	75W	180W	300W

These figures should only be considered as indicative ones - they may be used as average 'ideal' values for design purposes. Hence, for most water-lifting tasks making use of existing devices we can reasonably assume a human power output range of 30 to 50W. The lower value can be taken when considering sustained irrigation pumping, and the upper one as the maximum expected for short-time domestic water collection or for more 'all body' activity (excluding pedalling) of a more sustained nature.

Taking the 30 to 50W range and an overall (mechanical) efficiency for the water-lifting operation of 25 to 60% (the lower for some traditional methods and the upper for a good hand pump) then in terms of 'water watts'* lifted we may expect about 7 to 30WW. For sustained irrigation pumping using traditional methods the range is more likely to be in the 7 to 20 WW range.

*The term 'water watts' refers to the output power of the water actually lifted.

Having arrived at the most likely water watts, this can be related to the flow rate, Q (m³/h) of the device and the head, H(m) lifted by considering the potential energy of the water, i.e. (mass of water lifted X acceleration due to gravity X head) and so the 'power' of the water. Thus:

water watts (WW) = mass (rate) of water lifted (kg/s) X acceleration due to gravity (m/s²) X head (m)

$$= \frac{\text{flowrate } m^3/h}{3600} \times \text{density of water} \times g \times H$$

$$= \frac{Q}{3600} \times 1000 \times 9.81 \times H = 2.7^3 \, QH$$

(Note Q is in m³/h and H is in metres).

The Q-H relationship for different water watts, i.e. 75, 45 and 30W (corresponding to the typical adult power outputs quoted in Table 1), is shown in Figure 1.

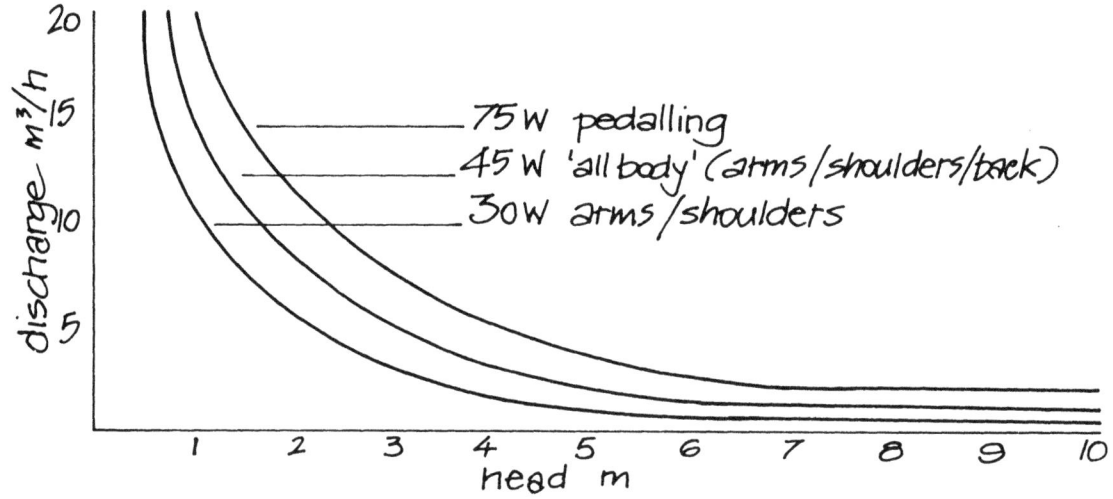

Figure 1. Q-H characteristics

The lines on the graph represent the impossible 'ideal' - a 100% efficient water-lifter/pump - for the different forms of sustained human-powered operation (i.e. mainly arms/shoulders; more 'all body' activity; and pedalling). Hence the curves represent the (unattainable) upper limit for the various forms of muscle groups used.

Figure 2 shows the Q-H performance data for various traditional human-powered water-lifters together with 10, 20 and 30W reference lines. Note that most of the devices fall within the 10 to 20WW range (7 to 15WW was predicted above). It will be appreciated that reliable data for these devices is extremely scarce.

The next section will give a brief description of these traditional devices.

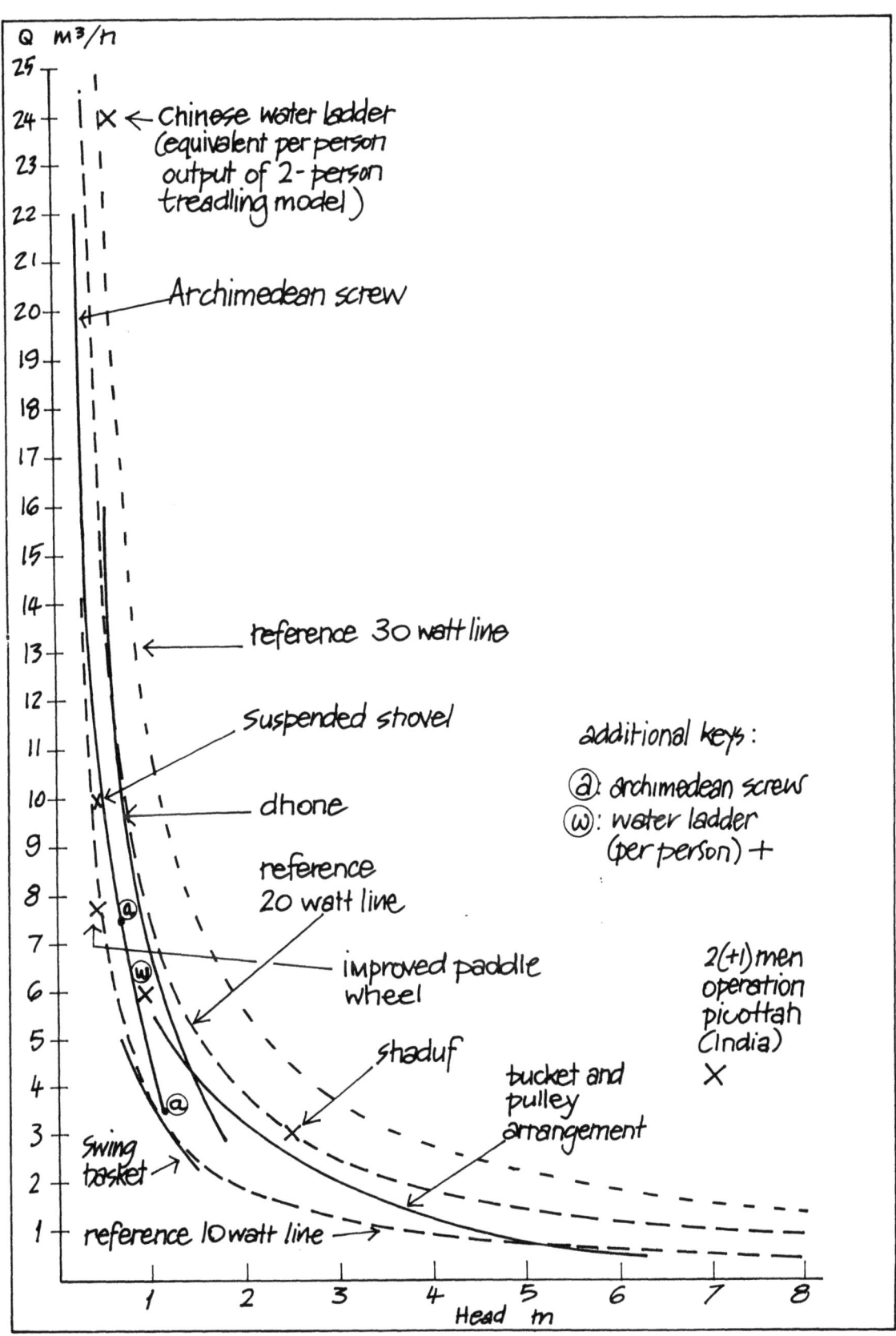

Figure 2. Q-H data for traditional devices

CAN YOU HELP WITH FURTHER DATA RELATED TO THE MEASUREMENT OF HUMAN POWER CAPACITY, OR THE WATER WATT OUTPUT, OF DEVICES? FIELD DATA (WITH ENVIRONMENTAL CONDITIONS INCLUDED) IS NEEDED.(SEE APPENDIX 1)

2.2 Review of traditional water-lifting devices

The first consideration in water-lifting methods is whether the water source itself is surface water (river, stream) or ground water, where the water-table has to be reached by a well/borehole. Hence the various devices may be classified as:
(a) surface water-lifters
(b) shallow-well water-lifters/pumps (up to about 6m lift)
(c) deep-well pumps.
Some devices may be used to lift either surface water or shallow-well ground water. In this chapter traditional surface water-lifters will be reviewed.

Traditional water-lifters include:
(a) water bowl; water scoop/shovel
(b) suspended shovel/scoop
(c) swing basket
(d) dhone (canoe-type pivoted trough/channel)
(e) paddle wheel
(f) water ladder
(g) Archimedean screw
(h) counterpoise lift (shaduf).
Note that the counterpoise lift is often also used for shallow-well water-lifting.

a) Waterbowl/jar

If the irrigated plot is less than, say, 200 m^2 (0.02 ha) water could be lifted and spread over a paddy field by a metal bowl. A child/old man can lift at the rate of 3 to 4m^3/hr from a water source to a field beside it, with a head of 0.15 m. This is a water watts output of 1 to 2 watts. This is certainly at the lowest end of the output scale.

Schioler (36) reports that at this end of the spectrum paddy fields are even irrigated by a (strong) man filling a 20-litre earthenware jar from a lake and spreading it over the field by using basins to throw the water. The low cost of US$0.60 (1979) is the only reason for using this inefficient, strenuous and time-consuming method. Hence the relative costs of the methods is a factor which has to be continually borne in mind when considering improvements.

A substitute for the bowls is a water scoop or shovel (Figure 3), which gives more of a sprinkling/'rain drop' effect. Schioler (36) notes that, 'the shovel should be light and formed so that it gives optimum efficiency/performance with regard to the shprinkling effect'.

b) <u>Suspended shovel/scoop</u>

Here the shovel is suspended from a tripod (Figure 4) which considerably eases the lifting task but makes the device less portable. The suspension is simple and ingenious - it is a double pendulum. It is used for paddy fields of less than 1 ha and for lifts of up to 0.5m with, say, 10 to 15m^3/h being lifted. It is naturally not very efficient. There is a suction zone behind the blade of the shovel which draws with it a great amount of water which does not reach the paddy field. Probably about one-third of the lifted water is wasted in this way. The overall efficiency is probably not more than 25%. The device can also be used to remove excess water from a flooded field. The total investment is of the order of a few US dollars (1979). The method is physically exhausting; ergonomically it involves only arms/shoulders and back muscles with some exaggerated body movement and it may need two men working in shifts to operate over a sustained period.

c) <u>Swing basket</u>

The swing basket (Figure 5a) operated by two people has the advantages of being cheap and mobile. It is used for lifts of up to 1m. Three sizes are commonly used, the smallest for a lift of about 1m, the middle-size (about 8 litres capacity but holding only 4 litres of water during the lift) for a lift of around 0.5m and the largest for a lift of 0.2m.

From an ergonomic viewpoint the operators work rather inefficiently due to the awkward movements required. They work at right angles to the direction of the bucket motion, bend, brace against horizontal and vertical forces whilst pulling and straightening up, then twist trunk and tip. The device is often worked on the basis of four men working in 2-hour shifts which allows water to be raised to a height of 0.6m at a rate of about 5m^3/h. The overall mechanical efficiency is likely to be only 25-30%. It appears that an irrigation area of up to 0.25ha (0.6 acre) of dry season paddy can be irrigated by swing basket for a lift of 0.2m. One installation in Bangladesh makes use of 8 operators, working in shifts with 2 swing baskets, to lift water in stages a height of 5.5m. The 'baskets' themselves may be made of wicker (Figure 5b, preferably lined with leather, or inner tube), or a tin bucket, or metal sheets, (Figure 5c).

Figure 3. Water scoop

Figure 4. Suspended scoop

Figure 5a. Swing basket

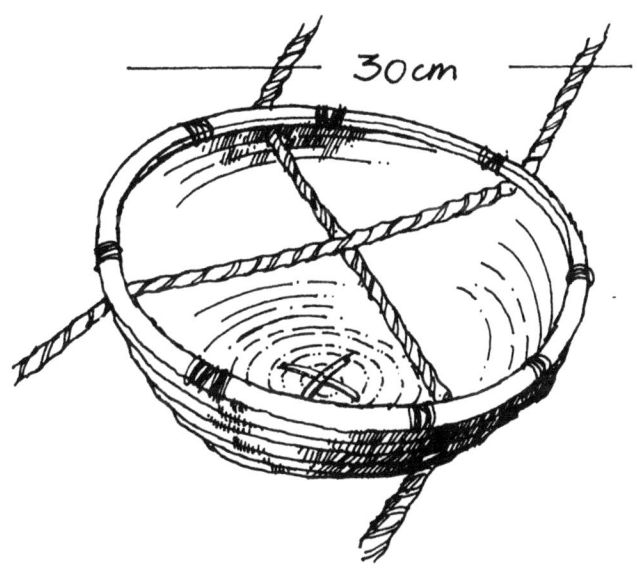

Figure 5b. Wicker swing basket of average capacity 8 litres.

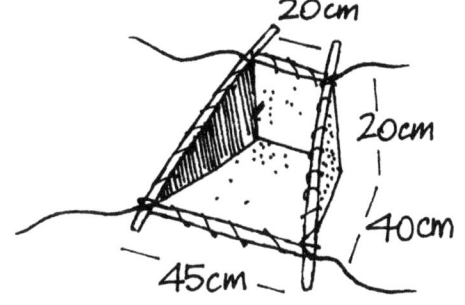

Figure 5c. Swing basket made from metal sheets

N.B. Not to same scale

d)   Dhone (canoe-type suspended channel)

The dhone (Figure 6a) consists of a water channel in the shape of a half-canoe made from a hollowed-out tree trunk, or assembled from planks. A straight canoe-shape can only lift water about half a metre whereas the curved type (Figure 6b) can lift to a height of 1.5m.

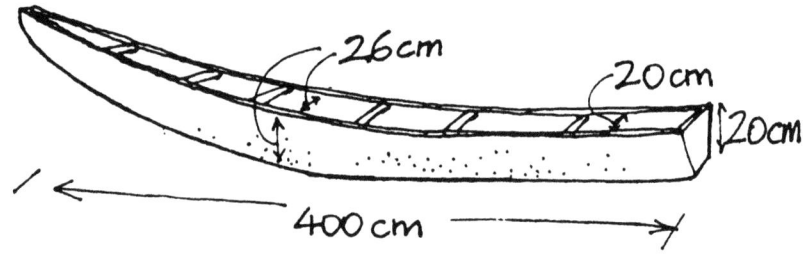

Figure 6a. Typical dimensions of a dhone channel

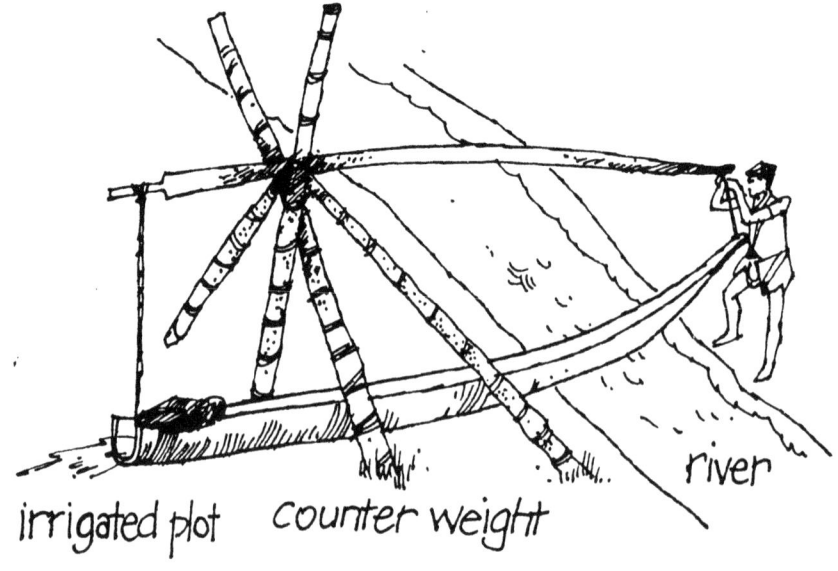

Figure 6b. The dhone in its upper position

The sequence of operations is: first, the person pulls down the upper beam connected to the canoe (a force of about 10kg is required). He then lowers half of the canoe beneath the water before transferring himself to the lower rod on which he stands. He then lifts the canoe until the counterweight takes over. At the stage where the channel is about empty of water he restrains the beam and stops the movement. About half of the potential (lifted) water never reaches the field since a large volume rushes back to the source at the moment the water channel comes to the surface. Although this method uses more 'all body' movements than some of the earlier methods considered, it is still strenuous and many rest periods are needed.

Note that as the farmer's land is normally scattered over several plots, it is important that he himself can move the canoe from field to field. In Bangladesh, Khan (21) has carried out some development work with dhones. Figure 7 gives details of Q-H characteristics of dhones used in both single-stage operation and with 2 to 3 stage operations (the latter used for higher lifts).

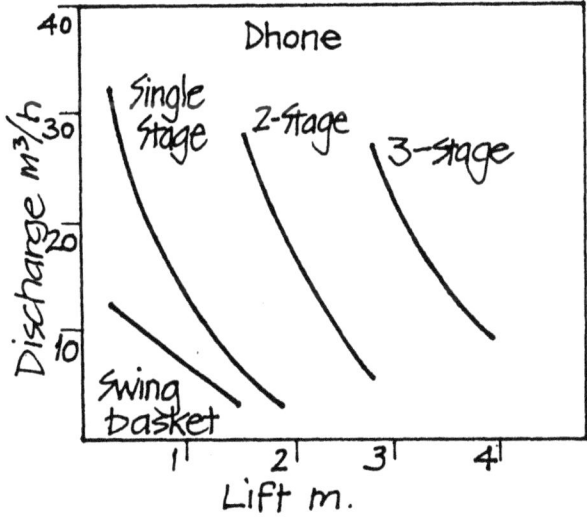

Figure 7. Comparison of swing basket and dhone

The dhone is widely used (probably in the region of one million) along the Ganges and Brahmaputra rivers in the Indian sub-continent. Some of the models used in these locations include curved ends. Khan (21) gives data which suggest a significant improvement in the potential irrigation area for dry-season paddy - a dhone giving about three times the irrigation area of that provided by a swing basket. However, in Bangladesh Khan has found that the higher initial costs of the dhone is the factor which inhibits its widespread adoption. There continues to be scope for some innovation in this method of lifting water.

e) <u>Paddle wheel - improved type</u>

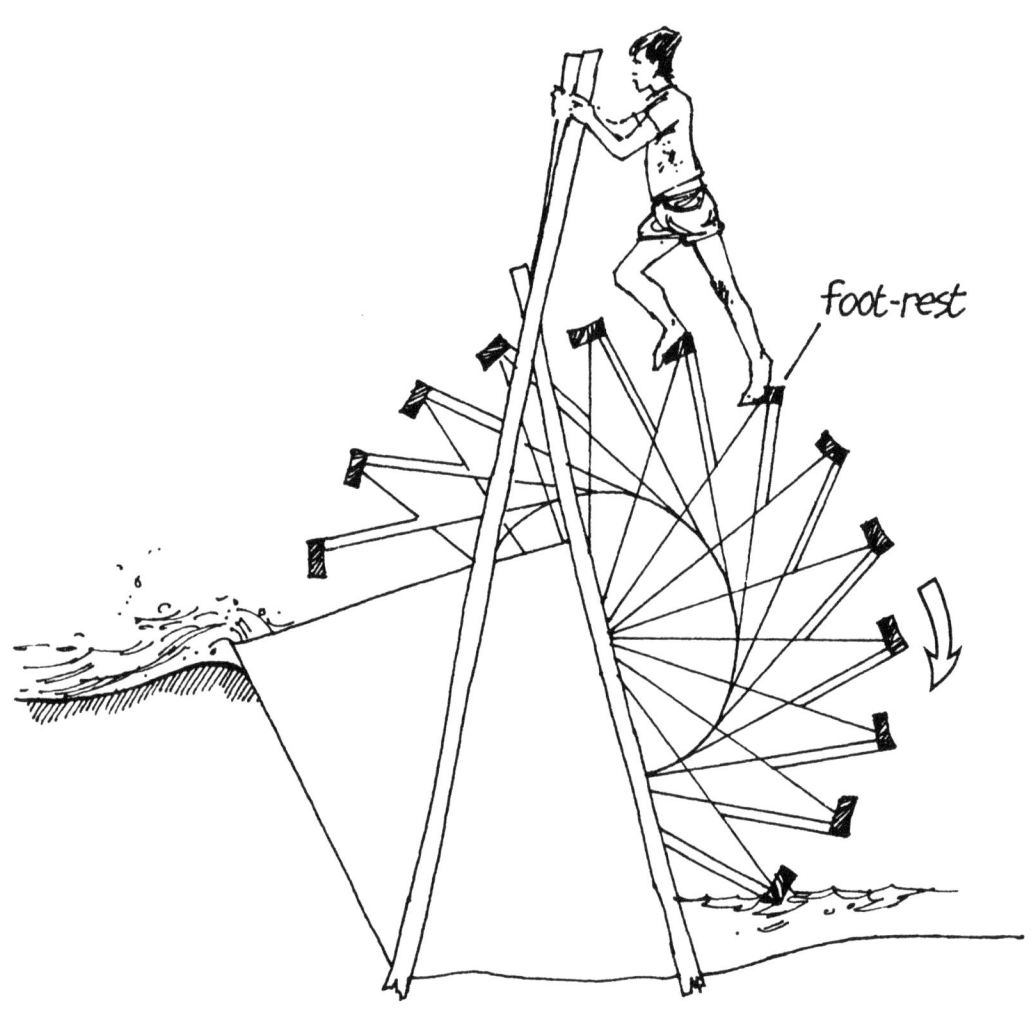

Figure 8. Paddle wheel - improved type

The original simple paddle wheel (an extremely inefficient device) is modified by placing the lower parts of the paddles in a close-fitting box (Figure 8). This reduces the spillage of water from the edges of the paddles so that an increased capacity is obtained or, alternatively, it is possible to lift water to a greater height. The operator powers the device by treading on the edges of the paddles, i.e. a tread-mill action. This is a low-lift device - certainly less than 1m lifts, but usually below 0.6m.

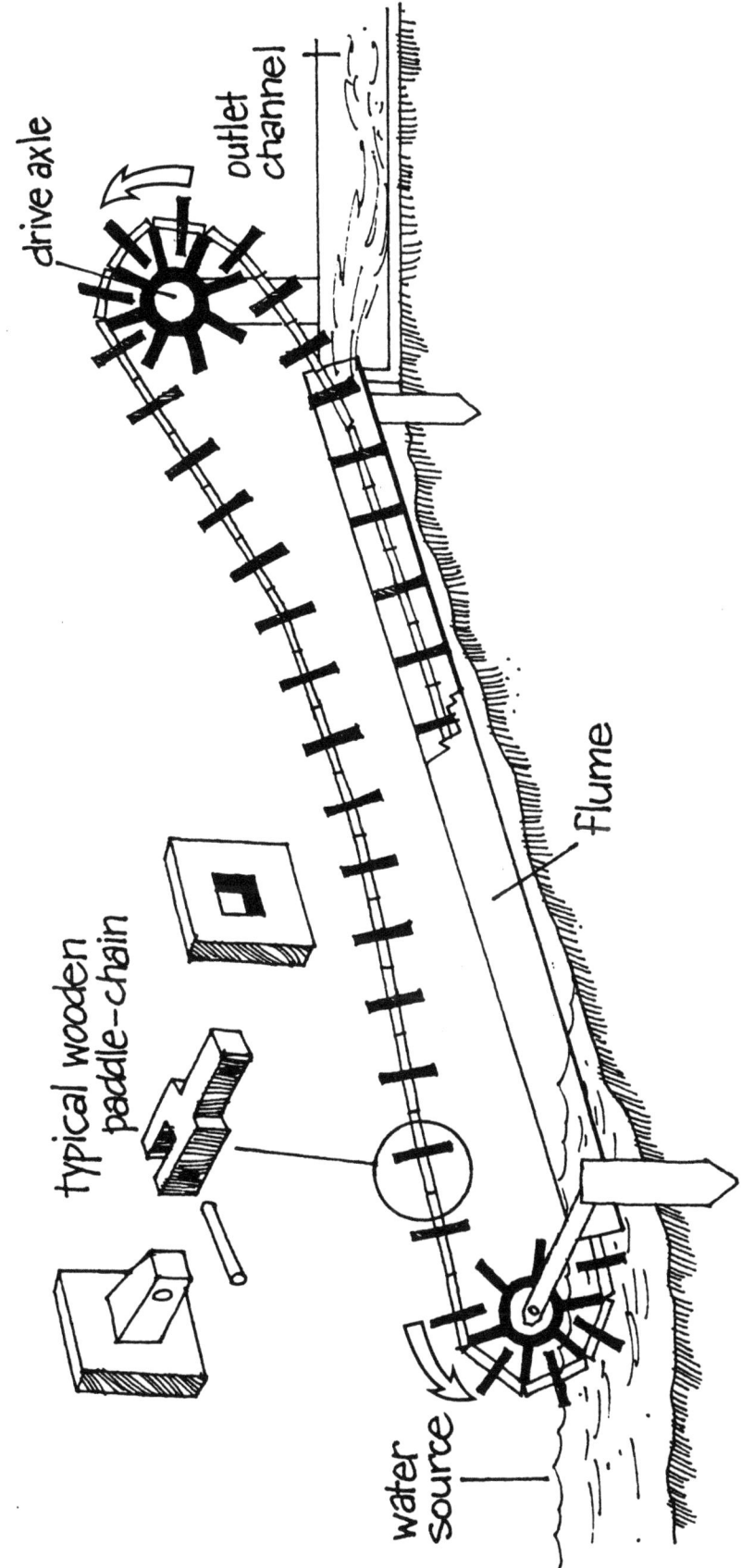

Figure 9. Water ladder

f) Water ladder (or Chinese dragon wheel)

This device (Figure 9) is used in China for lifting water from open streams. The original ladders were constructed almost entirely of wood. The device consists of a series of small wooden boards connected by wooden links to form an endless chain which runs through a tight-fitting wooden trough and around sprocket wheels at either end.

The human-powered water ladder may be classified into two types: hand-turning lift and foot-treadling lift. The former may be operated by a single person or by two together, and the latter by two or more persons. However, operation of the device is very arduous and draught animal power is used where available.

In present day China, improved (factory-made) water ladders (Chinese dragon wheels) are available with much better performance characteristics than the traditional all-wooden designs. Figure 10 shows the Q-H characteristics of two water ladders.

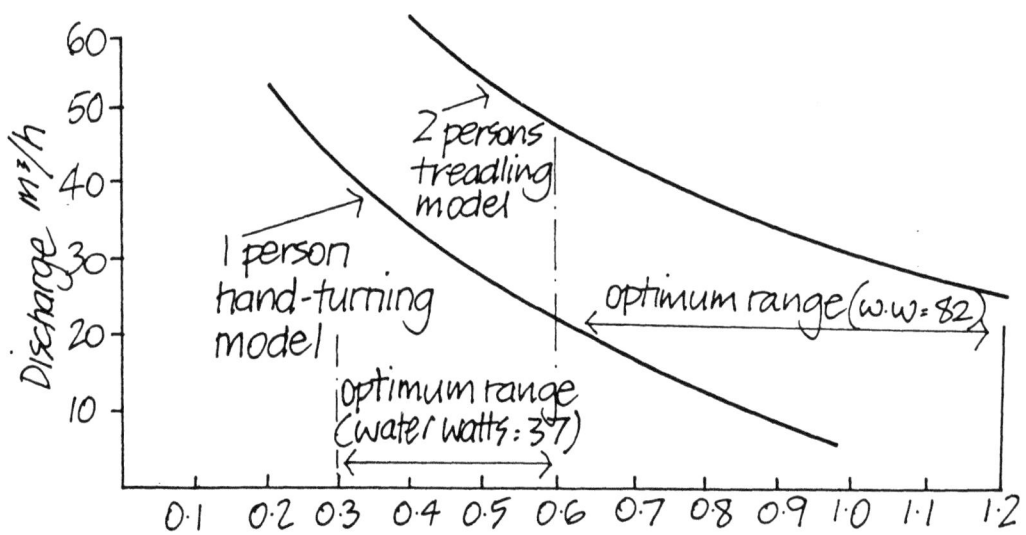

Figure 10. Q-H characteristics of Chinese water ladder

The traditional trough of the water lift is an open U-shape. About one quarter of the water is lost over the upper opening of the trough and through the gaps between the vanes and the inner walls. Contemporary devices have closed troughs with a rubber seal around the vanes to reduce leakage.

It has to be accepted that this device is fairly country-specific. As with other devices in China it has been part of traditional life and improved versions are still in use.

g) <u>Archimedean screw</u>

The Archimedean screw (Figure 11) consists of a wooden cylinder core with 'threads' of plank arranged as a helix around it. The assembly can rotate within a fixed hollow cylinder of the same length. During each revolution the lowest convolution of the helix picks up a certain quantity of water, and with continued rotation this water is gradually discharged at the top end.

Other forms of construction have been used in the past. Early literature refers to one or more flexible tubes (generally of lead or strong leather) wound on a cylinder of wood or iron. Instead of tubes wound round a cylinder, large grooves were sometimes formed on the wooden cylinder and covered by wooden boards or sheets of metal.

The device is usually hand-cranked by one or two men - either working together or in shifts. It lends itself to treadmill drive. This would solve the problem of hand-cranking which requires the operator's legs to be constantly immersed - giving rise to the problem of bilharzia which is rampant in the sluggish channels of Egypt.

The maximum length of the device is usually limited to about 3m which, with a maximum angle of inclination (usually about $30°$), means that it is limited to a lift of less than 1m (usually about 0.75m). Figure 12 shows the relationship between discharge and the angle of inclination of an Archimedean screw.

Laboratory tests (34) on an Archimedean screw show that while running at full capacity its hydraulic efficiency is approximately 75%. Figure 12b shows how efficiency is substantially maintained even when the device works well below its full capacity. At 25% of full capacity the efficiency is about 65%. Allowing for transmission and other losses for flow between 25% and 100% capacity, an Archimedean screw could have an overall efficiency in the range of 47% to 57%.

The pump delivery automatically adjusts to the inflow rate provided this does not exceed the 'fill-point level' (indicated on Figure 12b).

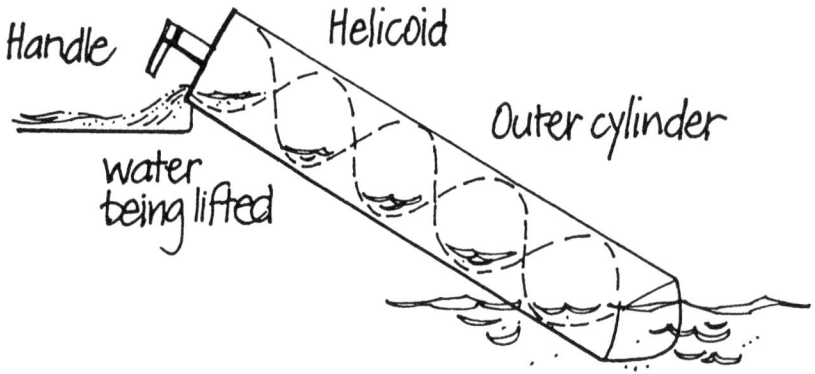

Figure 11. Archimedean screw

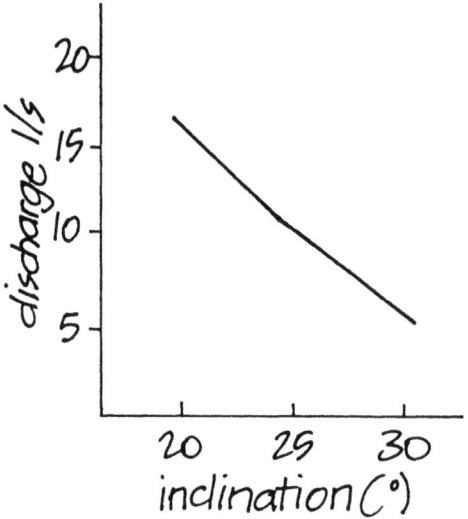

Figure 12a. Performance of Archimedean screw with uniform lift and speed

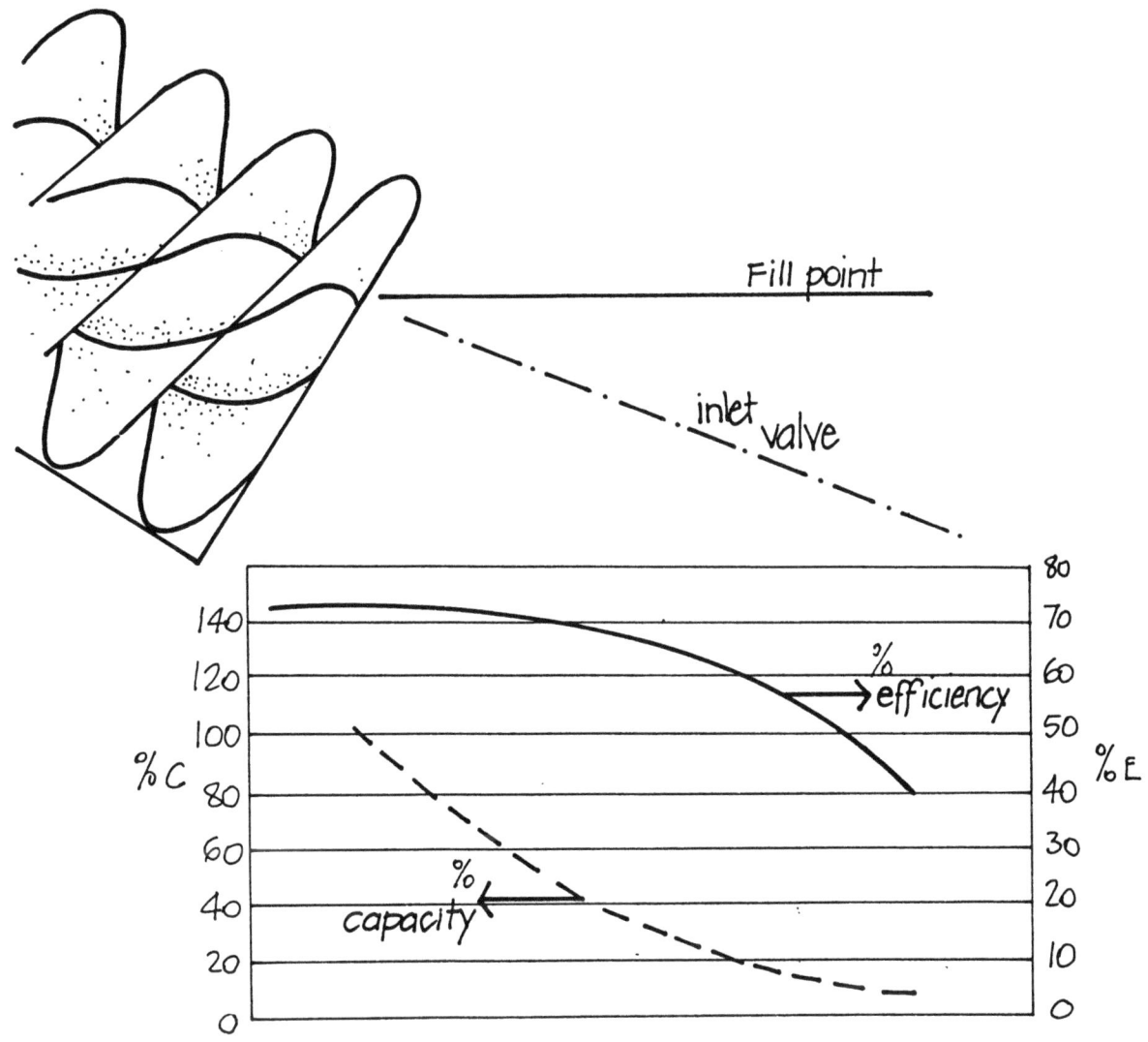

Figure 12b. Archimedean screw efficiency and capacity at varying inlet levels (Ref. 34)

h) **Counterpoise lift (shaduf)**

The counterpoise lift or shaduf (Figure 13) is an easily-built and easily-worked device (the best construction is found in Egypt). In its most common form it consists of a container (leather bag, petrol can, lined basket, metal bucket) attached to the end of a vertical pole which hangs from a lever with a counterweight on the other end. It is a device which uses the concept of mechanical advantage. The weight may be sufficient to balance half the weight of the full bucket so that the operator need only lift half the combined bucket and water weight. Then, to return the bucket to the water supply, he can use his body weight (to some extent) to affect the counterweight in order to pull the lever down. In many cases the counterweight mass may be such as to balance the entire bucket and water weight if the attendant is able to utilize enough of his weight to return the level. Any combination between these two extremes can be used depending on the lift, terrain and operator's liking.

Although the useful life of some of the materials used in its construction is short, the device is simple to maintain and can be readily and cheaply replaced with locally-available materials.

The picottah (Figure 14) is similar to the shaduf but larger and usually more massive. It is a high-lift counterpoise device which makes use of two persons on a moveable counterbalance beam (notched for footholes, and provided with a hand rail). A third person controls the water container/water flow. It can lift water to 7m, and is mostly to be found in India.

Here is an excellent 'body-weight shifting' device which eliminates much of the strenuous work load associated with the more arm/back muscular action lifting devices. Two hard-working men can have a maximum output of about 80 WW ($4.2m^3$/h @ 7m). The picottah is used primarily for lifting water from moderately deep-dug wells. The usual range of lifts is between 5 and 8 metres. With these higher lifts its output rate is low and so it is used largely to irrigate small vegetable plots. In Bangladesh a high lift counterpoise constructed of bamboo (Figure 15) which is very slender can lift water 6 to 8 metres. It is usually operated by a boy and can irrigate a garden of up to 0.04 ha. The output is about 20WW ($1.0m^3$/h @ 7m). These devices are made by the farmer himself and have many weak points (probably about 100 hours to breakdown), but they are fairly cheap and can be repaired quickly by the farmer.

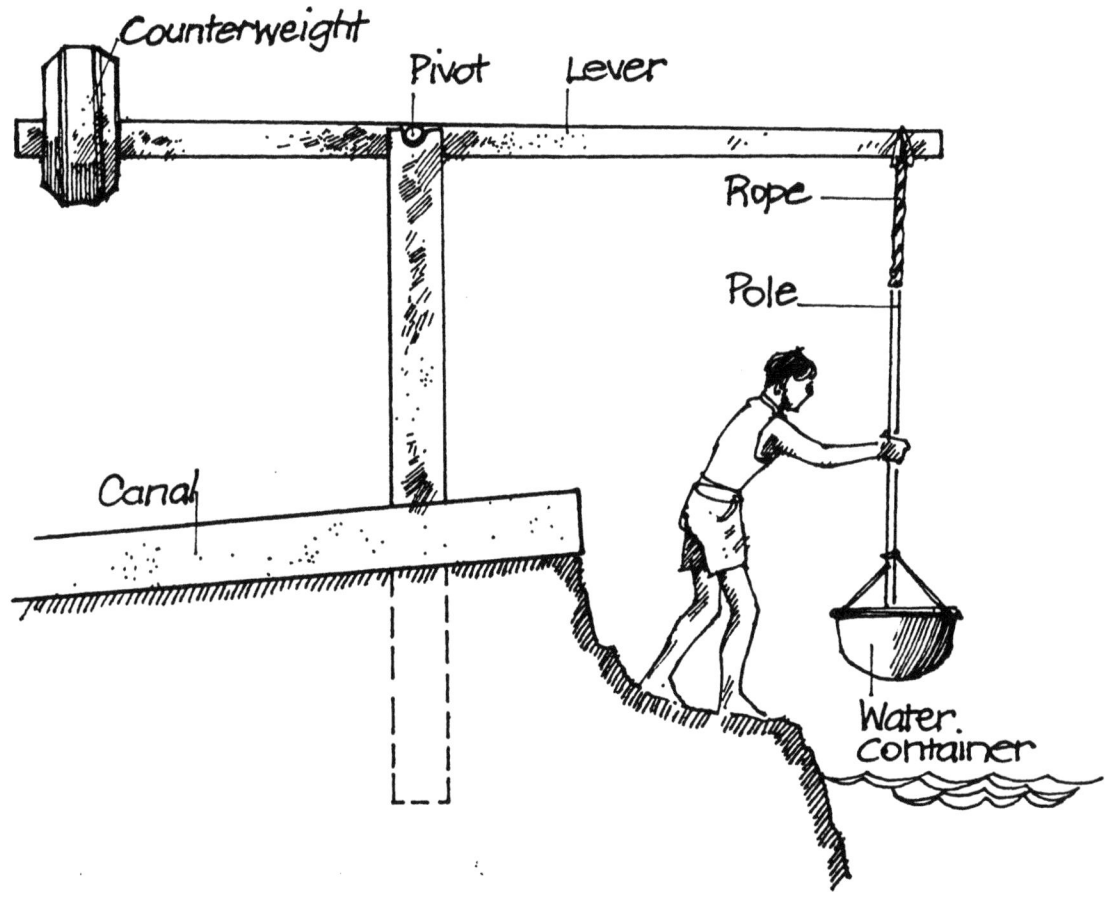

Figure 13. Counterpoise lift (shaduf)

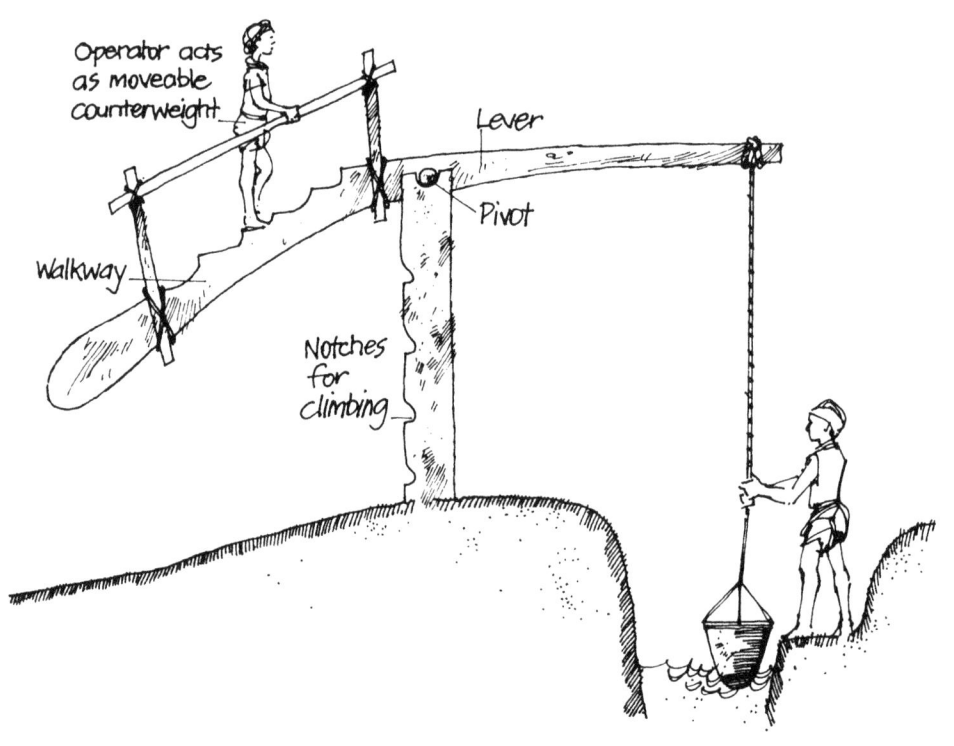

Figure 14. Picottah

CAN YOU HELP WITH FURTHER INFORMATION ON TRADITIONAL WATER-LIFTERS? WE NEED:

- FIELD DATA (Q-H CHARACTERISTICS), PRACTICABLE IRRIGATED AREA ETC.
- INFORMATION ON METHODS OF CONSTRUCTION, DIMENSIONS OF DEVICES, ERGONOMIC DATA ETC.
- INNOVATIONS ON THE BASIC DESIGN - ESPECIALLY THOSE WHICH HAVE EVOLVED FROM THE LOCAL KNOW-HOW OF THE OPERATORS, BUT ALSO OTHER IDEAS/PROTOTYPES (PROVEN OR OTHERWISE) OF 'OUTSIDERS'.
- INFORMATION RELATED TO SOCIO-ECONOMIC ASPECTS; THE EXTENT TO WHICH TRADITIONAL DEVICES ARE APPROPRIATE, ON THE DECLINE ETC.
  (SEE APPENDIX 1).

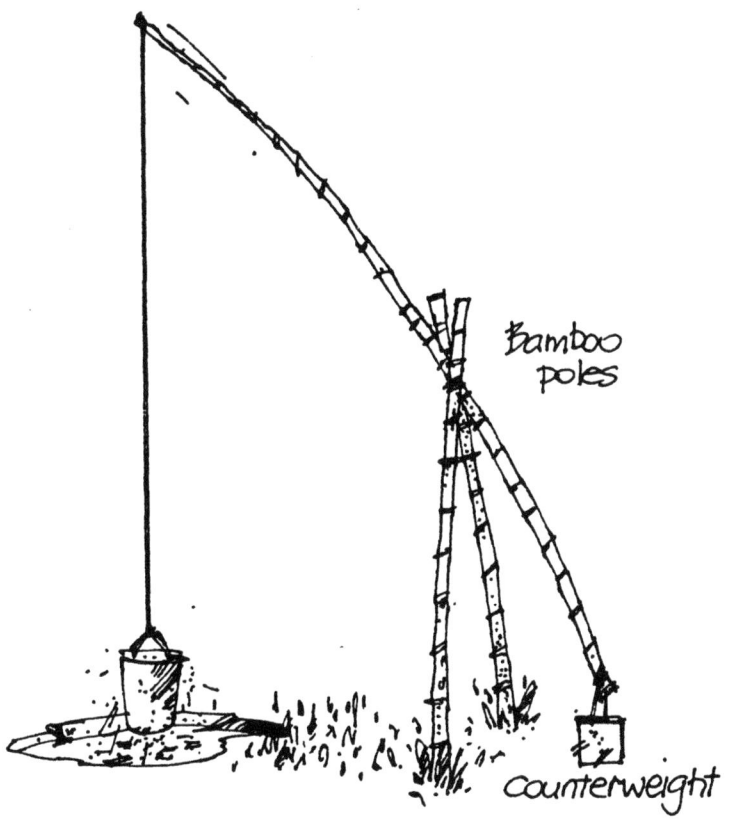

Figure 15. 7m high lift counterpoise for small garden
          (Bangladesh)

2.3 Practicable irrigated area of some traditional water-lifters

It is relevant here to give some indication of the irrigated area achievable in practice by the various devices which have been considered in the last section (Table 2).

Table 2. Practicable irrigated area of traditional water-lifters.

Device	Head (m)	Irrigation Area (ha)
Metal Bowl/Water Container	0.2 - 0.5	0.02 - 0.04
Suspended Shovel	0.5	0.35
Swing Basket	0.6	0.18
Swing Basket	0.2	0.25
Dhone	1.5	0.18
Dhone	0.6	0.4
Shaduf	2.5	0.1
Picottah (India)	7	0.16
High-Lift Bamboo Counterpoise (Bangladesh)	7	0.04
Archimedean Screw	0.5	0.4
Archimedean Screw	1.0	0.15

2.4 Research and development on traditional water-lifting devices

This section is intended to be an 'ideas generator'. We give some indications of possible developments in the area of traditional water-lifters. We hope that some of these ideas will 'take off' and that interested individuals and bodies will become involved in some real 'grassroot' research and development. We do, however, make the plea that developments in this area should be pursued with the fullest appreciation of the socio-economic realities and of the need to involve the operators of these devices in pre-project appraisal and on-going consultation. Once again we make a plea for further comment and ideas from the reader. To start the ball rolling here is a possible project:

<u>Field study and R and D on traditional water-lifting devices for irrigation</u> - especially manually operated devices.

The role and potential development of indigenous water-lifting devices - especially for irrigation - has really had no proper consideration in development thinking. If a 'bottom-up' approach to irrigation development is to be encouraged then a sensitive re-appraisal of the role of these devices should be made with the view to improving both hardware and operating methods. It is clear that there is scope for a research/development project in this area. Community participation would feature prominently in the field study, formulation of potential R and D, and in the resulting project implementation.

The field study part of the prospective project is essential, since at this 'lower end' of the appropriate technology hardware spectrum it is only too easy for an 'outsider' to suggest research and development possibilities which are not sensitive to the political and socio-economic setting of the community, and which may not respond to their 'felt needs' and indigenous know-how.

The field study would consider the type/particular design/materials/method of manufacture/cost/method of operation/ergonomic aspects/performance/usage pattern of existing devices. The political and social/cultural aspects of their use would be fully assessed, together with the degree to which they are on the decline and the interaction between traditional and other more modern water-lifting technologies in use locally.

The study would concentrate on a few countries and/or areas of a country where indications suggest that developments in the area of traditional water-lifting technologies would bear fruit. The study would aim to place the indigenous devices in the context of the demands/pressures of the present day. Questions like: 'Is there a felt need from the community to improve these devices?' 'To what extent have recent local innovations been incorporated?' 'What indigenous knowledge can be tapped?' would be answered. A sensitive re-appraisal of the actual situation is an essential pre-requisite to development of hardware and operational methodology and the objective would be to tap any intuitive know-how which the operators of these devices may already possess. To what extent are the operators able to adapt their device for ease of operation (e.g. how do they decide on the magnitude of the counter-balance weight on a shaduf?)

The study would consider to what extent there is scope to depart from human-powered devices. Is there a possibility of introducing (or expanding) animal-powered pumping technology, or simple, locally manufactured pumps? What are the bottle-necks to innovation/change?

The study would also consider the best way of promoting change, via community participation. Indeed community participation/'user-choice' strategy would be integral to the proposed field study stage in that the desires/needs and preferred choice of the community would be actively sought.

What devices might be considered in this project?

Some tentative suggestions for possible R and D are:
    (a) dhone (canoe-type suspended channel)
    (b) counterpoise lift devices (shaduf, picottah)
It is clear from Figure 2 that the dhone, shaduf and picottah appear to be the more efficient of the traditional human-powered devices which we have considered.

a) <u>Dhone developments</u>

In Bangladesh, Khan (21) has carried out a field study comparing dhones with swing baskets and other water-lifting methods (see Table 3). He has also suggested some modifications to the basic dhone design and its method of operation. Khan points out that in Bangladesh, since traditional methods of irrigation are used for watering about 40% of irrigated land, then if these methods could be improved there would be a significant increase in the area of cultivation or reduction in the time and effort spent in irrigating land. Khan considers that there is scope for improved operating methods - especially when the lift exceeds 1 metre, e.g. a two-stage lift system (two dhones operating in series), one lifting to, say, 0.75m and the other lifting from that height to the final desired elevation. Like all water-lifting devices the dhone suffers a reduction in volumetric efficiency at the upper range of its lift capacity; this is compounded by reduced overall (mechanical) efficiency of the human device combination due to impaired ergonomic factors associated with higher lifts. The overall result of these factors is a reduction in the water watts of the device of about 60% at a lift of 1.5m compared to a lift of 0.68m. In effect the outcome of this situation is that if two dhones are in operation discharging in series over a total head of 1.5m, then the discharge of this combination is more than 100% greater than if these two dhones were used side by side (in parallel) lifting to a head of 1.5m. There is obviously scope here for attempting to introduce improved methods of operation.

Improvements to the design of the dhone are also possible. In most cases the end which is lowered into the water is narrow and pointed. By widening and giving proper slopes to the bottom of the dhone at this end a greater volume of water could be collected.

Instead of using a treetrunk, a dhone may be made of several planks of wood which can be assembled in the field with bolts. Galvanized iron sheets can also be used. Very large dhones should not be used as Khan has shown that the discharge is reduced with increasing dhone volume (presumably because some of the input watts are wasted in manipulating the larger/heavier device (plus water) and so the time per cycle is disproportionately increased). Note also that it has already been observed that with increased lifts the overall efficiency of water lifting is considerably reduced. This is likely to be due to a greater spillage with increased lift, in addition to impaired ergonomics.

The possibility of incorporating flap-valves to aid the immersion of the dhone should also be considered (see Figures 16 and 17).

Figure 16 shows some further developments based on a gutter/channel water-conveying device which has some

TABLE 3: Comparison of Various Methods in Bangladesh (1979) - Ref (2)

Methods	Dhone	Swing Basket	Counterpoise with dug well	MOSTI[¥]	Lowlift Pump (LLP)	Deep Tubewell	Shallow Tubewell	Large Projects
Area Irrigated:								
In thousand acres	970	161	10	50	1 300	130	60	110
In percent (%)	35	6	0.3	1.8	46	4.7	2.2	4
Water Source: Surface/Groundwater	S	S	G	G	S	G	G	S
Source of Energy: Diesel/Electric/Manual	M	M	M	M	D/E	D/E	D/E	D/E
Discharge (m^3/h)	26[1]	8[1]	2	2.7	200	200	50	-
Pumping height (m)	0-1.5	0-1.5[2]	0-4.5	1.5-6	0-6	12 m+	0-6	0-10
Capital Cost (Tk)[3]	400	30	150	1200	35 000	180 000	23 000	-
Working Life (years)	4	2	3	6	7	10	7	-
Command Area (acres of dry season paddy)	4-5**	1-1.5**	0.33	0.5	40	40*	10*	-
Government Subsidy as % of Capital Cost	Nil	Nil	Nil	Nil	70%	90%	Nil	95%
Public Sector allocations in FFYP[4] (million Tk)	Nil	Nil	Nil	-	740	1 621	210	3 095
%	Nil	Nil	Nil	<1	13	28	4	54

[1] Maximum discharge
[2] Double stage systems can be used for higher lifts
[¥] Manually operated shallow tubewells for irrigation
[3] 15 Taka = 1 US dollar
[4] First five-year plan.

* Note: Taking a reasonable 'water-watts' for the dhone and swing basket of 20 and 7 WW, respectively, (see figure 2), then it can be calculated that the discharge of 26 m^3/h for the dhone is for a head of 0.25-0.3 m (perhaps its normal minimum working lift), and the discharge of 8 m^3/h for the swing basket is for a head of about 0.3 m.

** The irrigated 'command areas' quoted here for dry season irrigation of paddy seems to be very optimistic - see Table 2 where more realistic values for the swing basket and dhone are given. (Khan does note (2) that the potential areas are often not achieved since land holdings are often less than one acre).

similarity to the dhone, and which is used in Portugal (called cegonho). This design uses a weight-shifting principle and flap-valves - in the form of a see-saw gutter.

A variation of the dhone is shown in Figure 17. This arrangement allows some foot assistance to aid submergence of the device in the water. (As can be seen the level/counter-weight must be designed to provide a greater moment than a full gutter but less than the empty gutter plus operator).

A picottah-type dhone has also been suggested (Figure 18). Once again the ergonomic factors of the design should be given due consideration.

A sensitive field re-appraisal of the dhone would reveal to what extent changes in design and operating method would bring about improved performance and/or life. It is important in this assessment not to up-market the device to the extent that it is not possible for the poorest farmer to afford the improved design. (We must bear in mind that the main reason for farmers continuing to use the swing basket rather than the dhone is the greater cost of the latter!)

b) <u>Counterpoise lift</u>

The shaduf is perhaps the most familiar water-lifter in this category. In those countries where it is in widespread use its effectiveness is as much due to efficient operation - to such a degree that it could be called a fine art - as it is to the mechanism/design of the device.

There is a degree of variability in the magnitude of the counterweight with respect to the weight of the operator. The counterweight may be sufficient to balance half the weight of the full container so that the operator need only lift the same. Then, to return the bucket to the water supply the operator uses his body weight (to some extent) to counteract the counterweight in order to pull the lever down. In many cases the counterweight mass may be such as to balance the entire weight of the full container provided the attendant is able to utilize enough of his own weight to return the lever. Any combination between these two extremes is used in practice depending on the lift, terrain, operator's weight and individual preference. In this respect an arrangement utilizing optimum ergonomics is essential in order to maximize the output of the device.

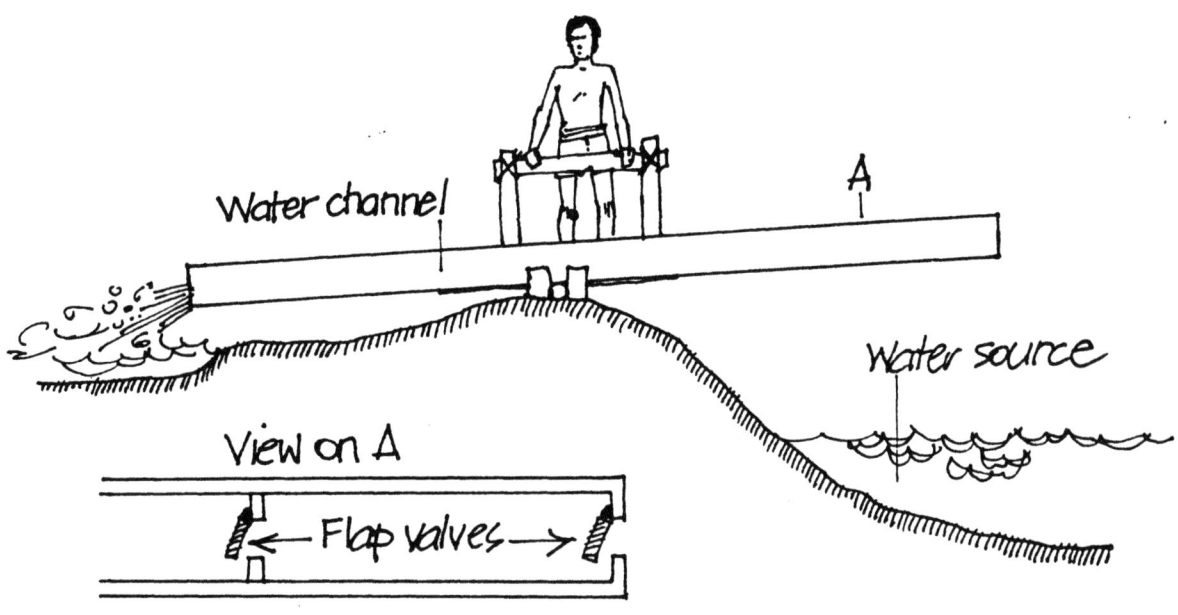

Figure 16a. See-saw dhone
Source: Wood, Ruff and Richardson (48)

Figure 16b. Inlet end modifications to increase
capacity of dhone
Source: Wood, Ruff and Richardson (48)

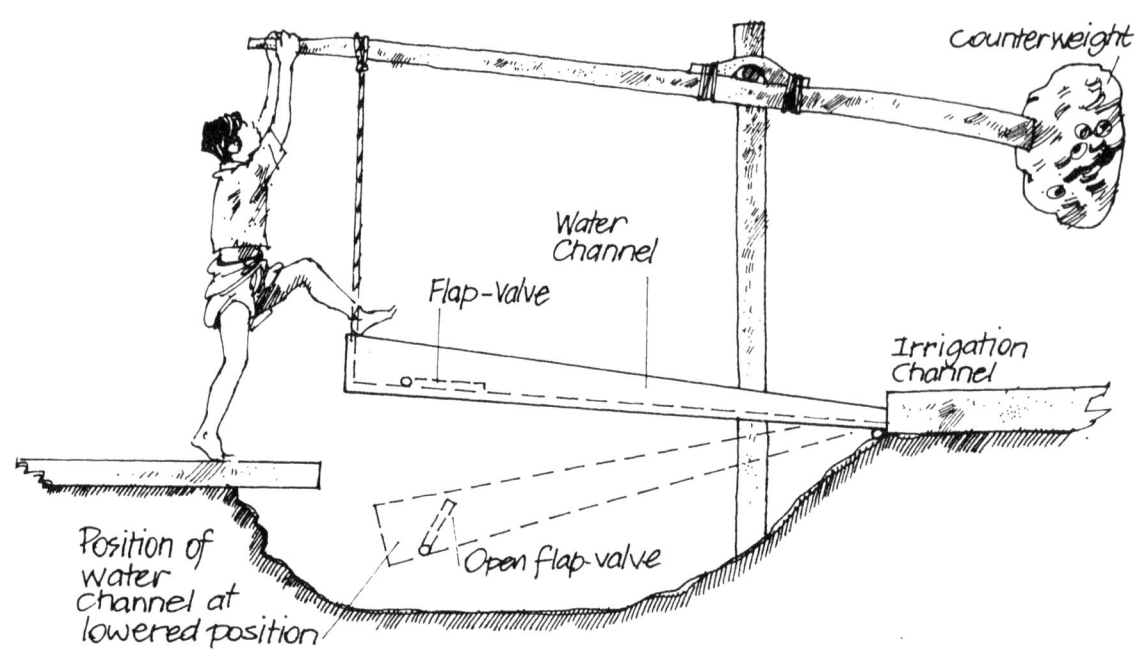

Figure 17. Dhone with flap-valve
Source: Wood, Ruff and Richardson (48)

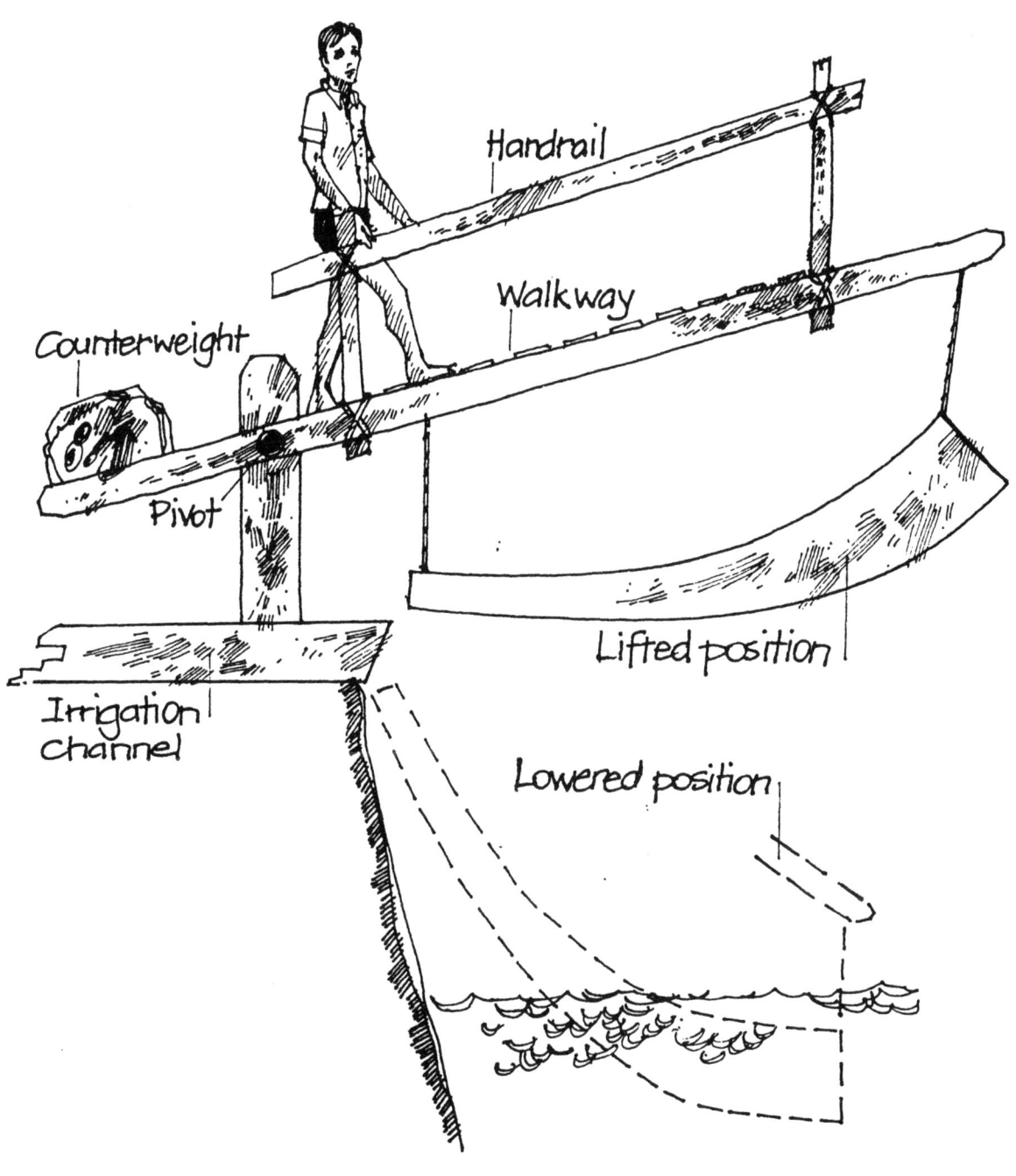

Figure 18. Picottah-style dhone
Source: Wood, Ruff and Richardson (48)

If the shaduf were to be introduced in a country where it has no previous use it would be essential to consider ergonomic aspects of the design and to optimize the operating method, i.e. to train operators in order to develop the 'fine art' associated with the use of the device.

One development of the basic shaduf is the picottah which is used in India and Bangladesh. The picottah can lift water to a height of about 7m whereas the shaduf is usually limited to lifts of about 3m. It is larger and more massive than the shaduf and in India two people are used as movable counterweight. The device is an excellent weight-shifting arrangement which eliminates much of the strenuous effort associated with the arms/back muscular action of many traditional devices.

There is scope here for considering to what extent a picottah-type device would find acceptance in those regions where the shaduf is more common - particularly where water has to be lifted to greater heights than normally attained. The introduction of a shaduf or picottah to those areas where it is not traditionally used could be a possibility, provided social factors related to its use are fully considered and provided that particular attention is paid to operating method. This will require a training programme.

DO YOU HAVE ANY COMMENTS ON THE SUGGESTIONS FOR R AND D OUTLINED IN THIS SECTION? IF SO CONTACT US (SEE APPENDIX 1).
IF YOU HAVE FIELD DATA - INCLUDING POTENTIAL IRRIGATION AREA FOR SPECIFIC DATA - WE WOULD LIKE TO HEAR FROM YOU.
FURTHER SUGGESTIONS OF POSSIBLE DEVELOPMENTS OF TRADITIONAL DEVICES ARE ALSO WELCOME.

2.5 Hand pump types

In this section the following hand pumps will be considered:
(a) piston-cylinder pump
(b) inertia (or joggle) pump
(c) diaphragm pump
(d) chain and washer pump

(a) Piston-cylinder hand pump

The piston-cylinder arrangement is the most common design found for hand pumps in most developing countries. Figure 19 shows the basic design, consisting of a piston with

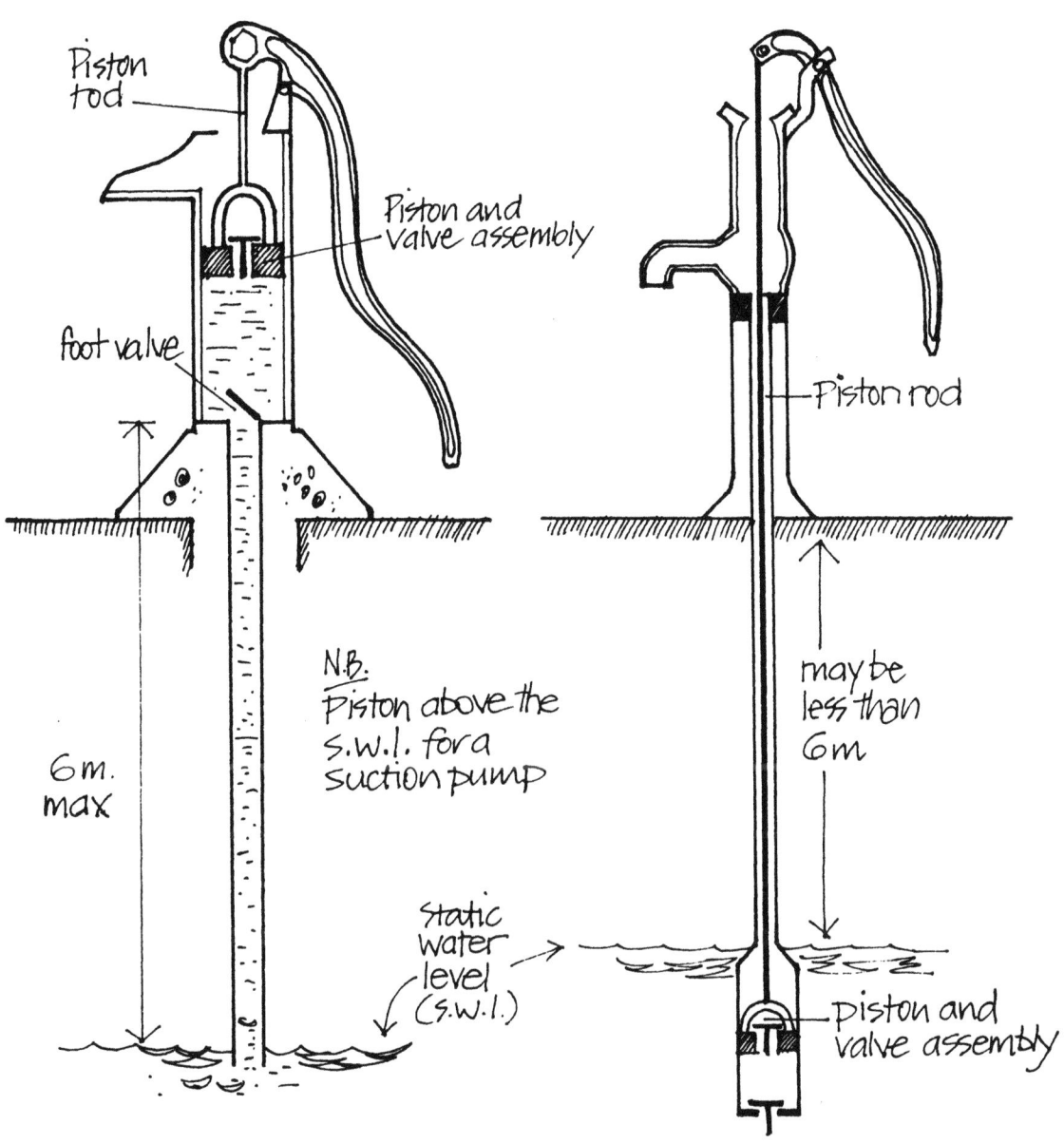

Figure 19a. Suction pump   Figure 19b. Lift pump

flap-valve and a foot-valve. This pump is also referred to as a 'suction' or 'lift pump'. As a 'suction pump' the piston will be above the static water level, provided the height to which the water is to be drawn does not exceed about 6m. In the non-suction lift mode the piston would be below the static water level. With this mode of operation the 6m maximum lift does not apply - the limitation now being on the maximum column of water which can physically be lifted by manual means.

The principle of operation is that with the pump primed, an upward movement of the piston creates a decrease in the water pressure under the piston and thus allows water to flow in through the foot-valve. On the down stroke with the foot-valve closed the piston forces water out through the piston-valve.

Since this basic design is considered to be the 'standard' design, individual manufacturers' pumps of this design will be considered in Section 2.7 (Hand Pump Testing).

The remainder of this section will consider some alternatives to the basic piston/cylinder design. The Q-H characteristics of these pumps are shown on Figure 20.

It has been found that several components of piston-cylinder pumps demand particular attention both at the design stage and in ongoing maintenance. The following notes give some comments on these:

Bearings:
For plain (bush) bearings any bearing surface should be adequate. The bearing length : shaft diameter ratio should not be less tha 2:1 (preferably nearer 3:1). In the case of wood block bearings these should be oil-impregnated (by hot-oil soak). These low-cost bearings are used on simple, low-cost pump designs (e.g. the proposed new deep-well Bangladesh pump - the Tara pump).

The elimination of bearing lubrication on many pump designs is achieved by the use of sealed ball races. However, these are susceptible to damage by impact loading and in many cases a self-lubricating bush would be a better solution.

Leather seals:
Leather piston seals can present considerable problems unless due care is given to material selection, manufacture, fitting and operation. The thickness of the leathers in relation to the clearance allowed is important - bearing in mind that leather swells when wet. The leather should not be too hard - the Bangladesh Tara pump is to have leathers which are chrome-tanned and impregnated with a non-toxic elastomer.

Unless due care is taken when they are fitted then jamming or difficulty in pumping can occur when the assembly has bedded in during the initial operating period. If the leathers are made and fitted correctly and are of the correct thickness, they are capable of giving a long life provided they are kept wet and the cylinder bore is smooth. If leathers are continually drying out and becoming wet again, rapid deterioration is bound to occur. In such cases some other seal material should be used.

Foot-valve seal:
It is important to have a good foot-valve seal to ensure that pump priming is not a problem.

Handle/lever:
In order to reduce wear at the top bush it is necessary to have a handle arrangement which allows for pump rod alignment variations due to handle movement. A three-pivot arrangement is needed to allow for this. Many pumps do not provide this - it being accepted that some wear can be tolerated for the sake of a simpler handle arrangement. Of course the simplest arrangement is to have no lever-handle - as with the proposed Bangladesh deep-set pump (Tara pump) which uses a simple direct-pull handle.

It has been found that, in pumps with a lever-type handle, a handle with a mechanical advantage of 4:1 is typical for shallow-well pumps and 8:1 (or 10:1) is advised for deep-well pumps.

Spout design:
A pump spout which slopes downwards will, to some extent, guard against objects being dropped down the riser pipe/cylinder.

External fittings/fastenings:
Ideally there should be only one external fitting - carefully protected and difficult to remove (dismantled with a special tool) - especially in those countries where pilferage is a problem.

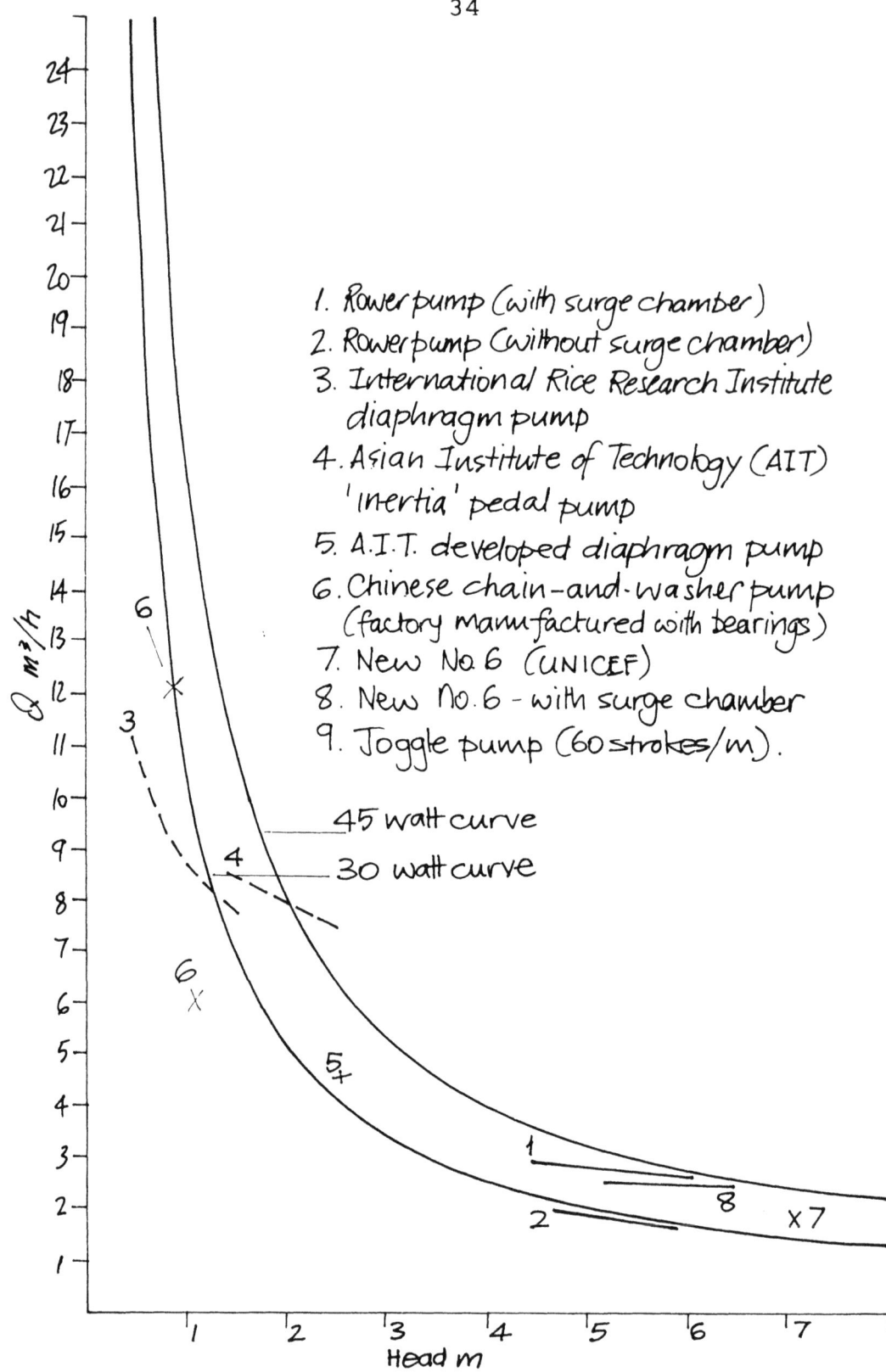

Figure 20. Performance of human-powered pumps

b) **Inertia (or joggle) pump**

The inertia pump is undoubtedly a very simple device for in its basic form it consists of a riser pipe with a flap-valve and a discharge spout (Figure 21).

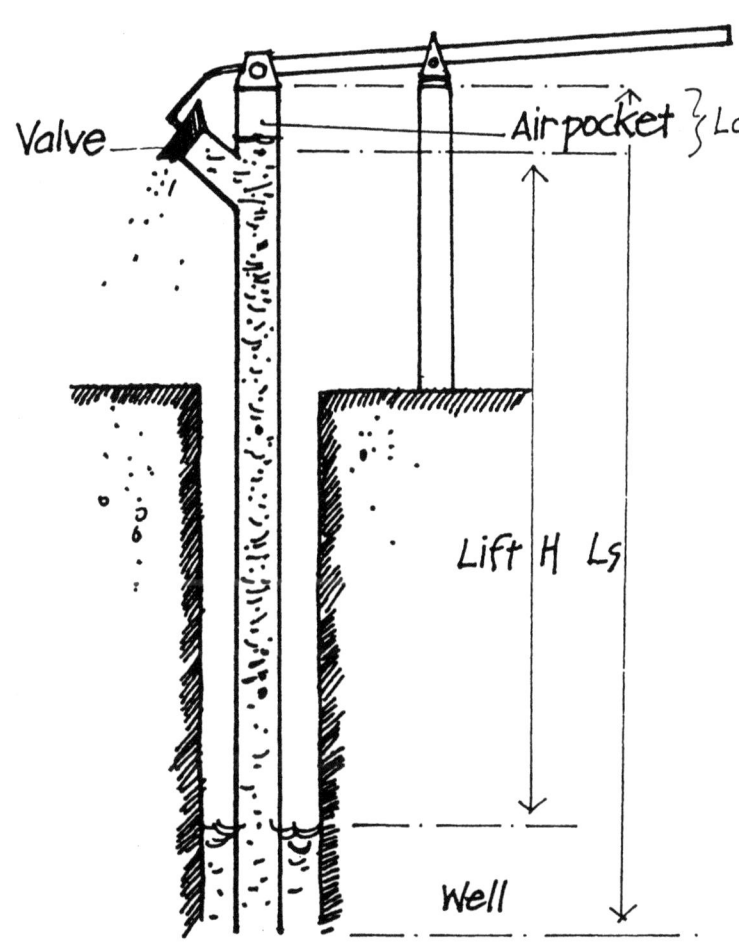

Figure 21. Joggle pump

The pump was initially called an 'inertia' pump since it was once thought that part of the basic principle behind its operation was due to the inertia of the mass of water held in the riser pipe. In fact, although the pump is remarkably simple in its construction, it does in fact rely on some quite complex flow mechanics - the induced flow principle. The principle of operation has been studied in detail by Burton (4) and the name 'joggle pump' is now used in preference.

Initially the tube is empty and has to be primed by maintaining the lower end of the pipe below the water surface and by rhythmic up-and-down movement of the tube. On each downward movement air is expelled from the closed tube via the flap-valve. At this point the pump is primed and ready for pumping water.

During pumping the operator needs to 'feel' the frequency-dependent nature of the pumping action. The water in the tube acts as an operating mass, whilst the air trapped below the flap-valve acts as a spring. In field tests it has been found that farmers have very little difficulty in acquiring the required 'feel' to operate the pump. Some practice is, of course, necessary.

Figure 22a and b give the performance details of avertically positioned pump and an inclined one (4). In Sierra Leone field testing was carried out with the inclined pump, because, during the dry season, gardens are planted in swamp areas where the water level is very close to the surface. Holes are dug 4 or 5 feet deep and the joggle pump is laid at an angle resting on the side of the pit with the lower end in the water.

<u>Tube length L = 2.74m; dia = 7.5cm; stroke 28cm</u>

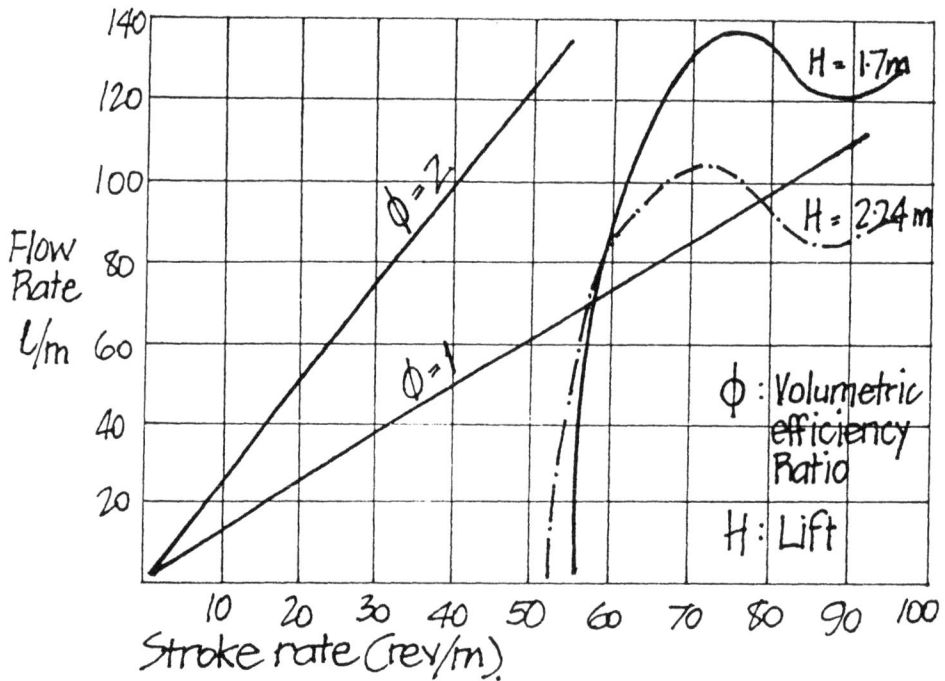

Figure 22a. Performance of vertical joggle pump

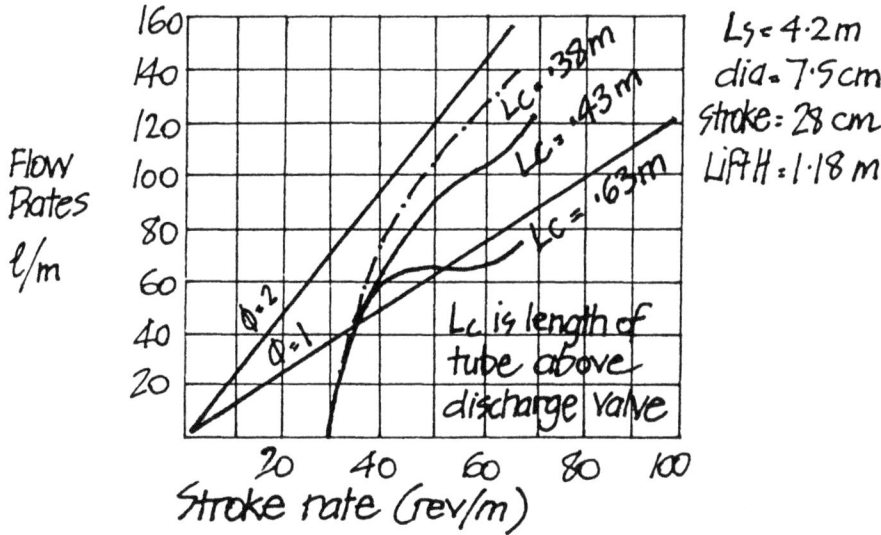

Figure 22b. Performance of inclined joggle pump

Joggle pumps are not easy to prime when the lift exceeds 5m, and it has been suggested that a foot-valve could solve this problem.

c) Diaphragm pump

The basic principle of this type of positive-displacement pump (see Figure 23) is that a suction effect is obtained when the pumping element - a flexible diaphragm - is lifted and so water is drawn in through the inlet valve. When the diaphragm is depressed, liquid is forced out through the delivery valve. The Vergnet pump and the Petropump (described in Section 2.7) are both based on the diaphragm principle.

The idea of producing a simple water-lifting device utilizing a pair of flexible bellows as the pumping element (6) was devised at the International Rice Research Institute (IRRI) in the Philippines, (18) where the prototype pump was developed for use in irrigation. Subsequently the Asian Institute of Technology (AIT Thailand, carried out a study on a modified version of the original design (39). The IRRI design was such that the pump body was required to be partially submerged during operation. The AIT design provided suction lines so that the pump body was free of the water. Both the IRRI and the AIT designs are operated by the farmer standing on the foot-rests and merely shifting his weight from one foot to the other - hence it is a weight-shifting type of human power operation. By alternately shifting his weight in a rhythmic manner, the operator is able to pump a continuous flow of water.

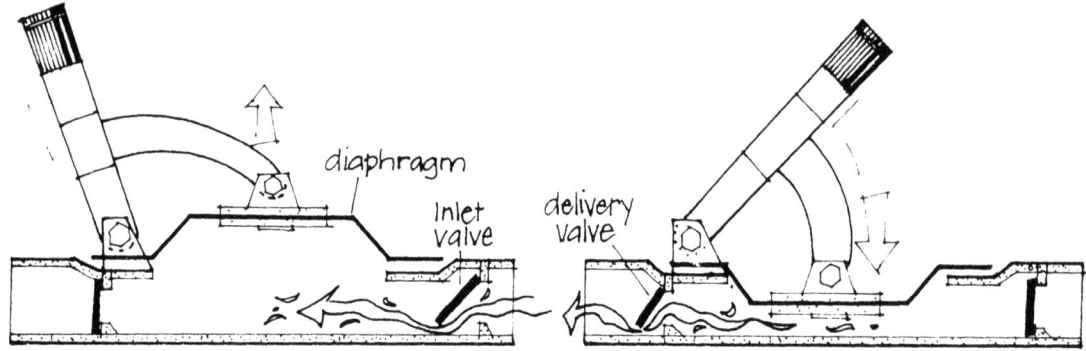

Figure 23. Diaphragm pump

d) <u>Chain and washer pump</u>

This pump (Figure 24) has been used for many centuries in China and Europe. It is still commercially manufactured - especially in China - and low-cost village-technology designs have been available (45). The principle of operation is simple: a continuous chain and washer disc loop is pulled up a riser pipe with a close fit between the washers and the pipe. The loop passes over a geared chain wheel, and down again to the bottom of the pipe. The bottom of the riser pipe is bell-mouthed to facilitate the entry of the washer/discs. There is a trough for water collection at the top of the riser pipe.

In China human-powered chain and washer pumps are available which are capable of lifting water to a height of 6m using one or two people (they are manufactured with a crank-handle on both sides) (12). Another version for use by one draught animal is capable of lifting water to a height of 12.5m (12). An overall efficiency of 76% and 68% respectively is claimed for these two units. It should be noted that these manufactured pumps have a metal chain wheel and metal roller bearings. At a village-technology manufactured level, with careful attention paid to shaping and fitting the washers, and with well-shaped chain wheel teeth and wooden bearings, an efficiency of around 50% should be possible. Figure 25 shows the general arrangement of a human-powered Chinese chain and washer pump.

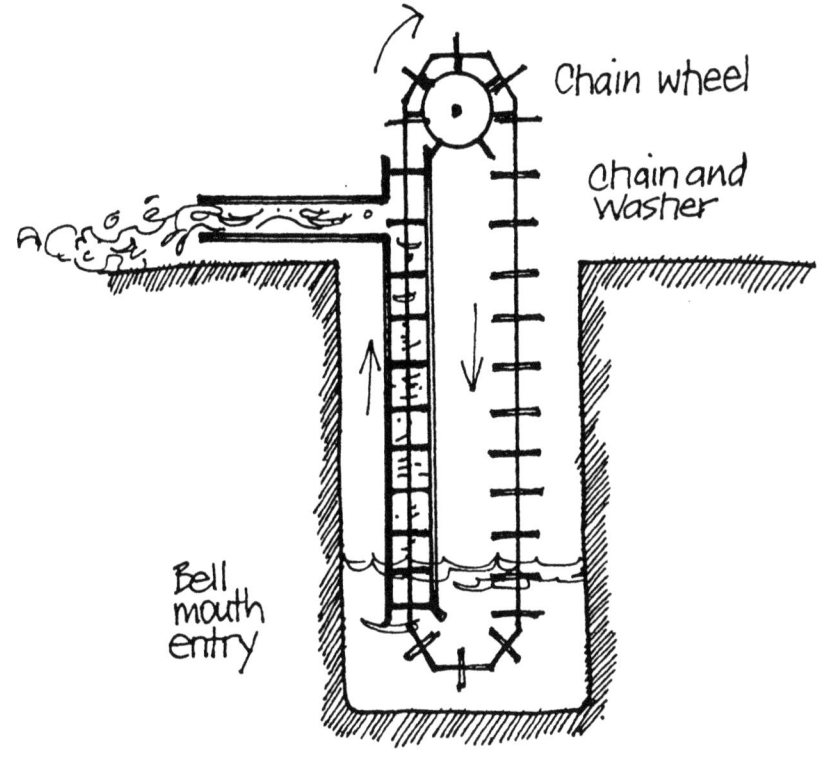

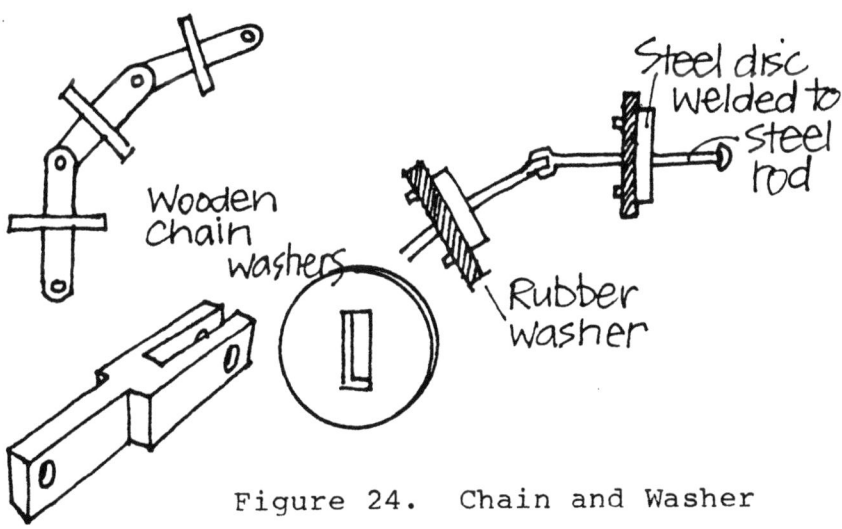

Figure 24. Chain and Washer

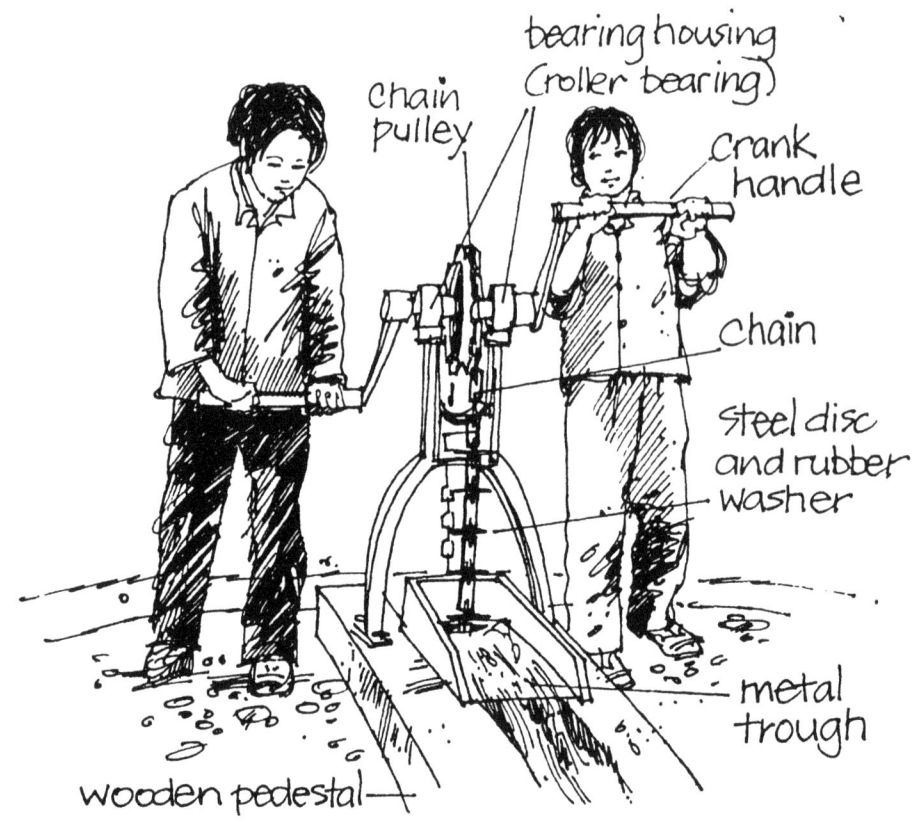

Chinese chain and washer pump

Figure 25

The simplest form of the Chinese pump (Figure 25) is suitable for lifting ground water from shallow wells and surface water from ponds or dykes. It is claimed that one or two men can use the device to lift 5 to 8m^3 per hour over a lifting head of up to 6 m. (Two men lifting 5m^3/hr at 6m head gives 82 water watts, about 40WW per person, which would suggest an input power of about 52WW, taking an overall efficiency of 76%, as quoted).

The main characteristics of the device are as follows:
- turning force required: 18kg (measured at handle, at 6m lifting head)
- rotational speed: about 34 rpm
- discharge rate: about 8.6m^3 per hour
- efficiency: about 76%
- working pipe diameter: 75mm
- outer diameter of circular rubber washer: 80mm.

2.6 Hand pump types: assessment/relative merits

Figure 20, shows the Q-H data for the pumps considered in this section. The Inertia (Joggle) pump is attractive in that it is the simplest design with the minimum of moving parts. It does, however, require a certain 'feel' in operating in order to get it to pump - and maintain pumping. It also has the disadvantage of not being easy to prime when the lift exceeds 5m (although this could be remedied by a foot-valve). In general it does not have as high an overall efficiency as the other hand pumps considered in this section (see Figure 20). A vertically positioned inertia/joggle pump gives a typical performance of about 20 water watts. This is no more than that attained by the better traditional water-lifters (see Figure 2). As a 'bottom-end' performance hand pump it should be considered in the same category as traditional water-lifters. That is not to say that it is not worthy of consideration in a particular situation. Acceptance by the users (in particular with the requirement for operator 'rapport') would be essential. Note that the water watts attained by this pump can be increased from 20 to about 25WW if the pump is inclined rather than vertical (see Figure 22). Note also (see Figure 20) that the Asian Institute of Technology developed a pedal-powered inertia pump which attained a performance in the range of 35 to 50WW. It may be that the difficulty of 'feeling' the best pumping would mitigate against the use of pedal-drive.

The diaphragm type of pump does not really offer any advantage over the piston-cylinder pump. Indeed as the working element is a flexible diaphragm which is deflected/stretched, the force required to operate it may be quite considerable compared with a piston-cylinder pump. This is borne out by the Consumers' Association laboratory test on the Petropump (see Section 2.7) which uses a diaphragmatic hose as the pumping element (8). This pump was operated, via a lever, by hand and it was not as easy to operate as other hand pumps. In addition the life of the flexible pumping element might well be limited. In the above CA tests on the Petropump the element failed.

The increased effort required to operate a diaphragm pump is presumably the reason why this device is usually foot-operated, i.e. the IRRI pump and the AIT-developed diaphragm pump, as also the Vergnet pump (the latter pump was tested by the Consumers' Association). Conversely the diaphragm pump lends itself - with its short stroke - to foot operation. The IRRI and AIT pumps are of similar design and operate by means of weight-shifting from foot to foot. The Vergnet pump (with a rubber diaphragmatic hose as the pumping element) operates by means of a foot-pedal action.

The chain and washer type pump is a proven low-technology pump dating back many centuries. However, to ensure a robust and dependable pump, care in manufacture is essential. The many links-washers assembly means that it is a multi-part pump and, as with any such device, dependability is a function of careful manufacture and assembly. The pump has been traditionally used in China, but the history and rural infrastructure of that country is far removed from the situation in many developing countries, especially in Africa. The inherent high friction with this type of pump - especially with deeper wells - makes it difficult to operate by hand with only one operator. The animal-driven chain pump is a good idea for those areas where draught animals are in use. The data in this section show that high efficiency units are possible - but note that these are good quality factory-produced units with metal parts and bearings. A two-person hand turning pump and an animal-operated pump, among others, are still produced in China (12).

FURTHER DATA TOGETHER WITH ANY RECENT DEVELOPMENT WORK ON ANY OF THE PUMPS MENTIONED IN THIS SECTION ARE MOST WELCOME (SEE APPENDIX 1).

UNDP/World Bank Hand pump Field Testing Programme

The UNDP/World Bank Hand Pump Field testing Programme started in 1981. It involves extensive field trials of over 2000 pumps in about fifteen countries. The pumps involved include those which have been found to be the most promising of those tested by the Consumer Association Testing and Research, UK (see Section 2.7)

One of the main objectives of the project is the development of 'village-level operation and maintenance' (VLOM) pumps which can be manufactured in developing countries and be able to be repaired by trained village operators.

Each test site includes 25 to 50 pumps, consisting of three to four different types. The pumps will be closely monitored and detailed data will be collected, analysed and disseminated.

A summary of each trial project in the various countries follows:

KENYA Funding agency: SIDA

Pumps to be tested: Maldev/Afridev, Petro, India Mark II, Prodorite (Blair).

The Maldev/Afridev pump developed in Malawi will be manufactured in Nairobi at an estimated cost of US$300.

TANZANIA Funding agency: FINNIDA

Pumps to be tested: Nira, Maldev, India Mk II, Prodorite (Blair), EZW-81(82)

MALAWI Funding agency: UNICEF, DANIDA

Pumps to be tested: Maldev, Consallen, India Mk II, Blair.

Malawi's test programme (US$400,000) will include further developments of the Maldev, including experimental plastic down-hole components.

SUDAN Funding agency: UNICEF

Pumps to be tested: India Mk II (experimental below-ground components), Blair, Maldev, Petro.

The Blair pump will be used on surface water schemes. The Maldev pump head has been designed to fit the India Mark II pedestal, and some will be introduced as alternatives to the Mark II's chain-and-quadrant arrangement.

GHANA Funding agency: Kreditanstalt fur Wiederaufbau (West Germany).

Pumps to be tested: Central Region - Moyno (USA), India Mk II (US cylinder);

Upper Region - Moyno (Canada); Monarch (Canada).

IVORY COAST Funding agency: World Bank; CIDA.

Pumps to be tested: Abi, Vergnet, Abi-Vergnet hybrid; Moyno, Monarch.

NIGER Funding agency: GTZ (West Germany), UNDP and others.

Pumps to be tested: India Mk II, Deplechin, Vergnet.

UPPER VOLTA Funding agency: Netherlands

Pumps to be tested: Volanta, Moyno, India Mk II, Vergnet, Abi-Vergnet hybrid.

MALI Pumps to be tested: Vergnet, Bourga, Abi, India Mk II, Abi Vergnet hybrid, Deplechin.

BANGLADESH The test programme will include a prototype deep-set handpump (the Tara pump), Rower (Bangladesh), New No. 6, India Mk II, Sarvodaya (Sri Lanka).

In all 500 pumps will be tested, mostly piston type and covering deep-well, non-suction low lift and suction mode. In Bangladesh the monitoring will also include a health impact study carried out in conjunction with the International Centre for Diarrhoeal Disease Research, Bangladesh

THAILAND Pumps to be tested: Korat 608, Dempster (modified), Lucky (suction), two local plastic pumps developed in Thailand through the UNDP country programme, Blair, Bangladesh deep-set (Tara), India Mk II, Maldev/Afridev.

PHILIPPINES Funding agency: World Bank

Pumps to be tested: locally made Jetmatic suction pump (based on Kawamoto), Blair, Local version of Clayton-Mark (intermediate and deep-set versions), Maldev head, India Mk II.

PAPUA NEW GUINEA Pumps to be tested: Blair plus others, including local experimental pumps developed by the PNG Appropriate Technology Development Institute.

CHINA It is hoped that the tests in China will be focussing on hand pumps for small-plot irrigation and will include some animal-powered devices. With the help of the UK Consumers' Association, two pump testing facilities are to be established.

LATIN AMERICA USAID have offered to operate field trials on projects under way in the Dominican Republic and Hondurus, and a forthcoming project in Peru. Pumps to be tested will include the AID/Battelle plus others.

2.7 Hand pump testing
2.7.1 Hand pump laboratory testing carried out by the Consumers' Association

An on-going laboratory testing project on hand/foot pumps has been undertaken by the Consumers' Association (CA). This work was initially funded by the British Government's Overseas Development Administration. To date, reports on the testing of four batches of pumps have been received from the Association (8).

It should be noted that the objective of the tests was to provide reliable performance, endurance and other data on the pumps with the hope that manufacturer's would be stimulated to make design changes. Many have responded to this challenge and continuing tests are being carried out on new improved designs. Hence the pump development 'scene' is not a static one - these tests have certainly made sure of that. The results given here are to some extent historical, as it is fairly difficult to obtain up-to-the-minute and reliable information (pump manufacturers may naturally claim that they have now improved their pumps - only independent testing verifies this).

It is beyond the scope of this report to make a full assessment of this ongoing development work. The reader should note that developments in hand pumps are to some extent in a 'state of flux'. Current water-pumping literature should be consulted for recent reports on this work.

The pumps tested at the CA Laboratories were of varied design and included some novel arrangements, including some which were foot-operated.

The first batch of pumps tested in 1978 were all deep-well pumps. They were:
    Petropump Type 95 - Sweden
    Vergnet Type AC2 (foot-operated) - France
    Dempster 23F - USA
    Mono ES30 (rotary operation) - UK
    Climax (rotary/flywheel type) - UK
    Godwin WIH 51 - UK
    Abi Type M - Ivory Coast
    GSW (Beatty) 1205 - Canada
    Monarch P3 - Canada
    Kangaroo (foot-operated) - Africa
    India Mk II - India
    Consallen LD5 - UK

The second batch of pumps tested (1981/82) was funded by the World Bank/UNDP. These pumps were:
    Korat 608 Al (deep-well) - Thailand
    Bandung (shallow-well) - Indonesia
    Briau Nepta (deep-well) - France
    Nira AF-76 (deep-well) - Finland
    New No.6 (shallow-well) - Bangladesh
    Moyno IV 2b (deep-well) - USA
    Kawamoto Dragon No.2 (shallow/deep-well conversion) - Japan
    IDRC Ethiopia Type BP (shallow-well) - Ethiopia
    VEW A18 (deep-well) - Austria
    AID/Battelle (deep-well) - Indonesia
    Atlas Copco (deep-well) - Kenya
    Jet-matic (deep-well) - Philippines

Batches 3 and 4 pumps were tested in 1983. These include:
    Two versions of the Volanta - Netherlands
    Maldev/Afridev - Malawi
    Modified Petro - Sweden
    Abi-Vergnet hybrid - Ivory Coast
    Funymaq - Honduras
    Sarvodaya - Sri Lanka
    Rower Irrigation Pump* - Bangladesh
    Two versions of Preussag - West Germany
    Two versions of Mono - UK
    RIHA - Australia

Reports on these tests were published in 1982/3/4 and a fourth is expected in 1985.

*Since a surge chamber is an integral part of the Rower pump, CA has also carried out tests on this component for UNDP.

The following brief notes give an indication of the merits/problems encountered with some of the pumps during the tests. (Details of the pumps may be obtained from the relevant Consumers' Association reports (8).

Petropump Type 95: An unusual diaphragm pump design using a hose as the pumping element. During laboratory testing the major fault which occurred was the failure of the pumping element.

Vergnet Hydropump Type AC2: This pump has a number of novel features - it is also a diaphragm-type pump which is foot-operated and has a hydraulic driving system. The main problem associated with this pump was related to priming (water was leaking from the hydraulic circuit).

Dempster 23F (CS): This is a 'standard' piston-cylinder pump design (see Table 4 for further comments).

Mono ES30: This is a rotary hand-operation pump. The operation of the pumping element is based on a kind of screw principle - a double-helical rotor rotates inside a triple-helical rubber stator.

Climax: Another 'standard' design which is well engineered but relatively expensive and of complex manufacture. It requires little maintenance but is very heavy (185kg in total for the pump stand) - hence installation and repair are more difficult than with other pumps.

Godwin WIH 51: A 'standard' design with a rotary hand-wheel operation. This pump was the most complex (and expensive) of the pumps tested in this batch.

India Mk II: A well-known design which has now been modified and considerably improved through experience in the field. It has a chain and quadrant system for maintaining pump rod alignment (see Figure 26).

This arrangement requires the weight of the pump rod to return the piston (on the down stroke). On deeper wells (ie: 20m) this proved satisfactory but it is likely to be a problem with shallower wells. This was one of the recommended pumps.

Consallen LD5: A simple 'standard' design which was one of the better pumps tested.

The World Bank has prepared a series of report forms for the field trials which collate information on pump installation, maintenance and repair, pump operation, pump deterioration, some well and water characteristics, and sociological/cultural conditions related to pump use. Two of these report forms are illustrated below: 'Site Inspection Report' and 'Breakdown Report'.

Site Inspection Report Form

To be completed periodically.

Where is the pump situated?

Name of village:

If the pump has a quantity monitor, what is its present reading?

What is the present water level?

Has there been any visible change in the pump?

       Yes? No?
       Wear
       Damage
       Missing Components
       If other, please give details

Describe wear.

Describe damage.

Identify missing components.

How many full strokes does it take to fill a standard container of x litres? (A full stroke is top to stop and return, or one complete revolution.)

Has there been any breakdown in the pump in the last month?

For how many days was the pump out of use?

What was the pump monitor reading when the breakdown occurred?

Have there been any other difficulties or problems with the pump or its use in the last month? Please describe as fully as possible.

Breakdown Report Form

If the pump has a quantity monitor what is its present reading?

What was the symptom of failure; what first went wrong which caused the caretaker to report a breakdown?

Which component(s) failed?

Which parts were replaced?

What is the distance from the top of the platform to the water surface?

Why did the breakdown occur?

       Mechanical failure?
       Damage?
       Missing components?
       Other?

How long was it from the time you began the repair to the time the pump was properly repaired and working again? Please write the number of days and/or hours.

Please estimate how many hours were spent specifically in repairing the pump.

Describe reasons for time taken between breakdown and final repair.

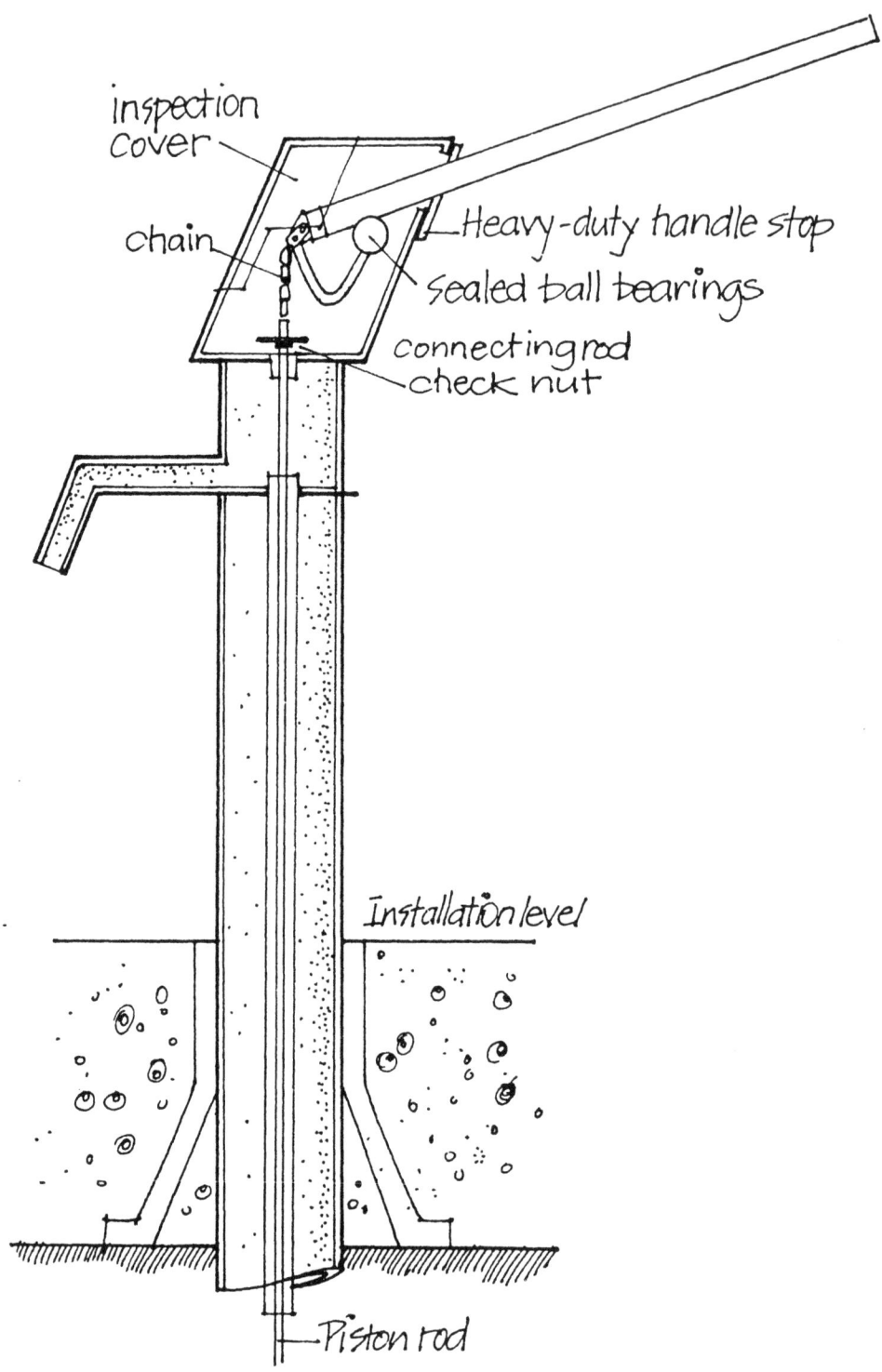

Figure 26. India Mk II pump

2.7.2 Results and conclusions of the Consumers' Association tests

Tables 4 and 5 give relevant test conclusions for Batch 1 and Batch 2 pumps respectively. From these it will be noted that:

(a) There is a great variation in the ex-factory costs of the various pumps. This reflects, to some extent, variations in the level of engineering and geographical cost.

(b) Of the Batch 1 pumps tested the two which gave the best overall performance-cum-manufacture/maintenance suitability were:

India Mk II deep-well pump (a very low-cost pump).
Consallen (LD5 model) (although some modifications required).

Of the Batch 2 pumps the best were:
Bangladesh New No.6 (shallow well)
Bandung (Indonesia) (shallow well)
Indonesia AID/Battelle (deep well)
Finland NIRA (deep well-but more expensive than others).

However, in general the Batch 2 pumps were not as good as the Batch 1 pumps, with the question of life-span being the main issue. It should, nevertheless, be noted that the endurance tests on many of the Batch 2 deep-well pumps were unrepresentative of actual field conditions since they were carried out with a power input which was 4 to 7 times greater than the normal range of human potential power input for the muscle groups used! (350 watts in at least one case!).

(c) Ergonomic aspects are of crucial importance, especially if the pump is to be considered for irrigation purposes. An ergonomic analysis has been carried out and will be discussed in the next section.

A technical report for the World Bank (51) covering Batches 1 to 3 pumps, noted that there were unsatisfactory features with all the pumps tested.
The Consumers Association did offer the following pumps as being worthy of consideration:

Shallow-well pumps
Ethiopia BP50: self-priming but not very robust.
New No.6: suitable for manufacture in developing countries provided foundry skills are available.
Rower: very suitable for low-lift, small-scale irrigation, but not recommended for drinking water supplies.

TABLE 4: Pump Assessment (Batch 1) - All Deep-Well Pumps

Pump (including country of origin)	Ease of Manufacture	Ease of Maintenance/repair	Contamination seal	Abuse suscept.	Safety Aspects	Ergonomic Aspects	Efficiency/Performance	Endurance Life	Cost* (Ex works) Estimated (at 1981/2)
Petro (Sweden)	complex welding required	problematic/ special tool needed	poor	poor	–	good and bad points	mechanical efficiency: low	pumping element has short life	$632 (excluding pipe)
Vergnet (France)	complex welding required	fairly simple	acceptable	reasonable	–	acceptable	priming problems fairly low mech. eff.	fairly good	$949 (excluding pipe) $1202 with 50 m pipe
Dempster (USA)	complex/ foundry required	frequent maintenance necessary	adequate	poor	–	fairly good	good	poor: excessive wear	$161 (excluding pipe and rod)
Mono (UK)	complex/ foundry required	not easy: infrequent maintenance (perhaps)	could be improved	no problem	–	fairly good	good	very good	$1054 (complete to 10 M)
Climax (UK)	complex/ foundry required	difficult: infrequent maintenance (perhaps)	poor		flywheel safety problem	good	good	handle failure but good otherwise	$1425 approx. (excluding pipe) $2083 with 21 m pipe
Godwin (UK)	complex/ foundry required	very heavy pump	could be improved	poor	–	not popular with users	adequate but low pumping rate	excellent	$2467 (complete to 21 m)
Abi (Ivory Coast)	fairly simple/ welding required	average	poor	poor	–	good	good	excellent	$1020 (excluding pipe and rod)
GSW Beatty (Canada)	Welding foundry required	frequent maintenance	–	poor	finger traps	average	fairly good	quite good	$465 (excuding pipe and rod)
Monarch (Canada)	foundry required	design problem	–	poor	–	average	good	quite good	$709 (excluding pipe) $1023 (with 30 m pipe)
Kangaroo (Holland)	Welding required	not worth considering	–	good	–	poor – not popular	poor	poor	$466 (excluding pipe and rod)
India MKII (India)	Fairly simple/ welding required	good	–	fair	–	good	good	excellent	$805 (with 20 m pipe) $188
Consallen (UK)	simple design/ welding required	OK with some modifications	–			average	quite good	good	$481 (excluding pipe and rod) $846 (with 20 m pipe)

* The large variation in price reflects variations in the level of engineering and geographical cost-of-living, etc. variations.

TABLE 5: Pump Assessment (Batch 2)

Pump (including country of origin)	Ease of manufacture	Ease of maintenance/ repair	Contam. seal	Abuse suscept.	Safety Aspects	Ergonomic Aspects	Efficiency Performance	Endurance Life	Cost ** (ex-factory) (As of 1981)
D Korat (Thailand)	fairly simple but foundry needed	good	fair	below average	quad and rack – bad finger trap	Handle too high; limited muscle groups	satisfactory	considerable wear: quad and rack	$295[I]
D Moyno (USA)	Complex/ high skill unsuitable dev. country	little maintenance needed but special tools	fair	good	sharp edges	awkward unpopular with users	output low with much effort required	no problems	$641[I]
D Briau Nepta (France)	Foundry/ forging welding needed	little likely but special tools	fair	average	finger traps	easy for children	very good	noisy bearings – handle pivot	$910[I]
S New No.6 (Bangladesh)	very simple	very easy	poor	poor	sharp edges	very good	very good	marked det. in performance after 1000 hours	$20(?) supplied free UNICEF ex. pipe/rods
D Nira (Finland)	fairly complex foundry etc.	not easy special tools	fair	average	sharp edges	much effort required	much effort required	handle broke 314 hours	$330[I]
S Bandung (Indonesia)	foundry welding required	fairly easy	average*	poor	finger trap	very good	very good	rubber cup seal split 800 hours	$54 ex. pipe/rods
D Kawamoto Dragon No.2	average/ foundry needed	difficult	poor	poor	finger trap	awkward	average/ poor below 7 m	–	$184 ex. pipe/rods
D Atlas/Copco (Kenya)	foundry/ forging welding needed	easy	poor/ average	poor	sharp edge	awkward	average	severe wear*	$669 ex. pipe/rods
S IRDC (Ethiopia)	simple	very easy	good	average/ poor	finger trap	awkward/ difficult	poor	–	$75 ex. pipe-rods
D VEW A18 (Austria)	complex but no foundry	very difficult	poor/ average	average/ poor	counter- weight hazard	very difficult to operate	average	severe wear* plus broken handle	$1286[‡]
D Jetmatic (Phillipines)	average/ foundry	average	poor	poor	finger trap	awkward	good at higher heads	dislocated cylinder	$38 ex. pipe/rods
D AID/ Battelle (Indonesia)	Foundry but other- wise simple	average	poor/ average	average/ poor	finger trap	very good	good	severe wear* plus con. rod broken	$120 ex. pipe/rods

D Deep-well pump
S Shallow-well pump

[I] Price includes drop pipe/rods
♀ Not surprising since the input watts was 200 (a factor 4-7 times greater than normal human input power).

[‡] Price includes operating cable/counterweight.
** The large variation in price reflects variations in the level of engineering and geographical cost-of-living, etc. variations

* Similarly high input watts.

2.7.3 Ergonomic aspects of pumps

The Consumers' Association laboratory tests on pumps included some U.K. citizen 'user' trials involving 11-year old children, women and men, (60 in total) of various statures (8). These trials on each pump consisted of the users being asked to fill a 10-litre bucket and answer questions about the height and comfort of the handle, the effort required and the overall ease of use. (An initial period of familiarization with each pump was allowed).

In addition to the 'questionnaire approach', objective observation of the users was also carried out. The shallow-well pumps were operated at a head of 7m. The deep-well pumps were operated at a simulated head of 20m.

These user tests are most informative as they reveal the importance of good ergonomics in pump design. The CA reports (8) do not analyze in depth the link between the ergonomic aspects of the pumps and their performance. From the data on performance (users' comments plus power-driven tests) and the discussion on pump ergonomics in the CA reports it is possible to analyze the relationship between these. Table 6 shows the results of this exercise.

The main conclusions of this study are:
(a) As the CA reports point out, it is important for optimum performance that the user makes use of an 'all-body' movement - utilizing as many muscle groups as possible: arms, shoulders, back and legs, without exaggerated body movements. Those pumps which allowed arm/shoulder movement only were more tiring and sustained pumping (more than the 10-litre test) would be exhausting.
(b) A comparison between the users' tests and the power-driven pump tests gives some indication of the human watts input to the pumps. In the cases of the shallow-well pumps for the two which allowed many muscle groups to be brought into play, (Bangladesh New No. 6 and Bandung) an input of 55W (on average) for 20s was sufficient to fill the 10-litre bucket. In the case of a deep-well pump which allowed many muscle groups to be used (AID/Battelle) an input of 100W for 30s was sustained without complaint from the users. This latter outcome is important as it shows, fairly conclusively, the degree to which a good body movement increases the power output of a human.

In contrast to the good body movement with a power input of 100W for 30s (with the AID/Battelle pump), the more erratic input effort (see comments on Table 6) required of the VEW A18 pump (with rotary-type handle) giving an estimated input (mean) of 114W for 31s was unsatisfactory and very difficult to operate. ALso the Nira AF-76 pump was found to be difficult to operate because of the high levels of effort required even with an input power of 85W for 35s, simply because a full body movement was not possible. Note that 100W is tolerable (for at least 30s) with the involvement of

TABLE 6: Comments Analyses of (Laboratory) 'User' Trials : Ergonomic Aspects

Pump	Type	Handle Type	Head (m)	Q(m³/h)	Water Watts	Input Watts	Pumping time (10:1)	Comments
Bangladesh New No.6	shallow	Lever type (MA = 4.7)	7	1.81	34.5	52	20 s	Good user response. Handle movement allows <u>arms, shoulders, back and legs</u> to contribute.
Bandung	shallow	Lever type (MA = 5)	7	1.76	33.5	48	20 s	Handle allows <u>many muscle groups</u> to contribute.
IRDC (Ethiopia)	shallow	Spade-type handle no lever	7	0.74	14	"40"*	50 s	'No lever' arrangement with considerable difficulty for small children. Most of the effort has to be supplied by <u>arms and shoulders</u> only.
VEW (Austria)	deep	rotary-type handle (MA = 1)	20	1.15	63	"114"*	31 s	Difficult pump to use with awkward interaction between pump and user except for those with sufficient strength. Most found it difficult to keep the handle turning smoothly as a result of being unable to 'time' their efforts on the handle.
AID/Battelle	deep	lever-type (MA = 5.8)	20	1.21	66	101	30 s	Most users operated pump without difficulty. <u>Many muscle groups</u> could be called into play without <u>exaggerated body movements</u>.
Briau SA (France)	deep	lever-type (MA = 7.1)	20	0.64	35	"46"*	56 s	Low handle compounded by other difficulties. The handle moved through a wide arc and most users tried for a full stroke because of the pump's slow rate of delivery but found the <u>exaggerated body movement uncomfortable</u>.
Moyno IV	deep	rotary (two hand grip)	20	0.45	25	"80"*	80 s	Disliked. Effort supplied by <u>arms and shoulders</u> only. The efforts were <u>high</u> and rate of delivery slow. Smaller users, particularly children, could not maintain a smooth circular motion of the handle.
Korat 608	deep	lever-type (MA = 10.8:1)	20	0.75	41	"79"*	48 s	Many users complained handle too high. Handle is long but stroke is relatively short. Users found it difficult to bring several muscle groups into play; most of effort by <u>arms and shoulders</u> only
Mica AF-76	deep	lever-type (MA = 5:1)	20	1.02	56	85	35 s	Children and small women found this pump difficult because of the high levels of effort required.
Kawamoto (Japan)	deep	lever-type (MA = 7:8)	20	0.68	37	"74"*	53 s	Many users found it difficult to decide on the best method of operation for this pump due to large arc of handle movement with consequent <u>exaggerated</u> body movement for many users who chose a full stroke.
Jetmatic	deep	lever-type (MA = 7:8)	20	0.44	24	"40"*	82 s	Similar to Kawamoto pump.

* Note that these input watts values are <u>net</u> values (derived by a comparison with power-driven tests), they are not a relevant indication of the <u>gross output</u> of the person for these pumps with poor ergonomics/exaggerated/limited body movements.

many muscle groups, but 85W for about the same time is not tolerable for a pump with restricted body movements.

Another interesting example is the IDRC Ethiopia shallow-well pump. In this pump a lever-type handle is not used - the handle is a 'garden spade' type. Here an estimated net pump input of 40W (in practice the gross ergonomic design) for 50s is difficult to achieve due partly to the poor ergonomics and partly to there being no 'mechanical advantage' assistance.

It is obvious from studying these comments that ergonomic aspects are rarely considered in pump design - only 3 (2 shallow, 1 deep) of the 12 pumps in the series 2 batch are considered satisfactory from an ergonomics viewpoint.

2.8 Hand pump research and development

2.8.1 Various prototype pumps

(a) The prototype Malawi pump (MALDEV pump)*

In Malawi a new approach to the problem of rural ground water supplies has been developed based on matching a better understanding of the geology of the area and the occurrence of ground water with the most appropriate and economic method of abstraction (42). Up to 1980, the 'rule of thumb' in searching for ground water was to drill into fresh rock - down to 50m in places Hydrogeological investigations revealed a means of bringing about a radical change in this method. It was found than an extensive shallow, weathered rock aquifer was present throughout much of Malawi, into which high-yielding boreholes up to 20m deep could be bored. This meant that heavy, expensive drilling equipment could be dispensed with in favour of lightweight 'percussion' rigs which were much cheaper and could be towed by small vehicles.

With Malawi's successful experience in community participation in its gravity-fed piped water programme, the same involvement is sought - for the first time - in the present integrated ground water programme. This has involved the development of a suitable hand pump.

The Malawi pump's most important feature is the facility for connecting-rod, piston, foot-valve and rising main to be removed by hand through the pump head. (The previous arrangement required a truck-mounted winch to lift the pump head off its foundation). With the new pump the possibility of village-level maintenance is opening up. The Malawi hand pump (Figure 27) has a steel well casing - a pipe with two steel brackets welded on. The brackets contain two sealed bearings. A bolt passing through the bearing centres attaches the pump handle to the body and provides a pivot for the handle. A T-piece at the end of the handle facilitates pumping.

*This pump features in the UNDP/World Bank field trials.

in this section show that high efficiency units are possible - but note that these are good quality factory-produced units with metal parts and bearings. A two-person hand turning pump and an animal-operated pump, among others, are still produced in China(9).

Desirable Hand-Pump Characteristics

1. **Acceptable engineering design**
   The design should be simple - with the minimum number of (preferably standard) parts. Particular attention should be paid to adequate hearing surface, lever/hand-wheel mechanism and ease of maintenance.

2. **Ease of manufacture**
   To facilitate manufacture in developing countries, manufacture should not involve complex processes and machining - depending on the country, foundry and forging facilities may not be available so that fabricated body work with welding/brazing may be more acceptable.

3. **Minimum/ease of maintenance**
   Maintenance requirements should (ideally) be as few as possible - compatible with simplicity of design and bearing in mind the capital cost of the pump. Here one has to balance the (possibly) conflicting requirements of simplicity, initial costs and maintenance costs. Maintenance/repair should be easily carried out with simple tools.

4. **Minimum initial/ongoing costs**
   A pump with a low _initial_ cost may not necessarily be the best buy. It may be a false economy to over-simplify a design in order to reduce the initial cost. It should be borne in mind that many water programmes have maintenance costs over a few years far in excess of the initial cost of the pump. A cost assessment must take into account all costs over, say, a five-year period.

5. **Reliability**
   Reliability may be defined in terms of the probability of the pump giving satisfactory performance without failure over a specified period of time (with minimal maintenance).

6. **Acceptable performance**
   The pump must be easy to operate and deliver an adequate discharge. The volumetric efficiency and mechanical efficiency should be as high as possible. The performance should not deteriorate significantly with worn parts and/or between maintenance checks.

7. **Acceptable to users**
   The pump should be acceptable to users and be fully compatible with their cultural traditions. In addition ease of operation - with reliability and acceptable performance - are important. Basic ergonomic aspects should be satisfied.

8. **Resistance to abuse**
   The pump should be resistant to abuse - accidental or by vandals - and the pilferage of parts should be made as difficult as possible. It should be robust (compatible with reasonable weight).

9. **Material compatibility**
   The materials used in manufacture should be compatible with the environment and usage rate and the water quality (corrosive, sand, etc.)

10. **Water contamination resistant seal**
    It is important that the pump is installed on its base with a proper seal. (It is also important to ensure that there is a concrete 'apron' forming the base of the pump installation, with adequate drainage.)

India Mk II pump

The India Mk II pump (Figure 26) is a deep tubewell hand pump evolved by UNICEF, India and a group of voluntary agencies in Maharashtra state. The pump uses welded steel fabrication, fabrication, sealed lubrication roller bearings, and it has a single pivot linkage connecting the handle (with an 8:1 mechanical advantage) to the pump rod through a 'chain' and quadrant arrangement. This arrangement facilitates pump rod alignment during pumping. The pump cylinder and piston are both cast iron. The cylinder is brass-lined and the piston has a first grade leather bucket seal.

The original design criteria included the following:

(a) The pump must have a trouble free life of at least one year, serving possibly 2000 people and drawing water from a static water level (S.W.L.) of 45m. (In fact the average installation serves 500 people with an S.W.L. of approximately 20m.)

(b) The design must be suited to local manufacture; not require any imported items or material; and be able to be maintained effectively by personnel having a minimum of engineering skill.

(c) Ease of operation must be a prime consideration; one adult must be able to operate the pump without difficulty. (To facilitate ease of operation, the handle - with a mechanical advantage of 8:1 - is of solid construction and is designed to counterbalance the pump rod weight at a depth of 20 to 25m. It has been found that the pump can be operated by a ten year old child.)

(d) The 'above ground' components must ensure hygienic standards - the pump head mounting to the platform must prevent any chance of pollution to the tubewell by surface water intrusion.

The mass production of this unit in India must not be under-estimated, since there is a high order of accuracy required in its manufacture. Initially, an attempt was made to utilise small-scale manufacturers in the country and establish production in each Indian state. However, it was found that basic engineering skills did not exist generally within this sector. In order to establish a reliable production central manufacture was established in Madras.

With the production of a satisfactory heavy-duty pump head, an attempt was made to develop a durable cylinder using plastic materials. Initial problems were encountered with regard to the quality control of the plastic materials on a mass produced level. However investigations in this area are continuing.

With the provision of a durable hand pump it was essential to ensure both an adequate standard of installation and the creation of a viable maintenance structure. Unlike a shallow-well hand pump (which allows village-level maintenance), this deep-well unit requires recovery equipment. This is done by mobile maintenance units at district level. A three-tier maintenance structure (district level; block level and village level) has been set up in some Indian states (see Section 6.3).

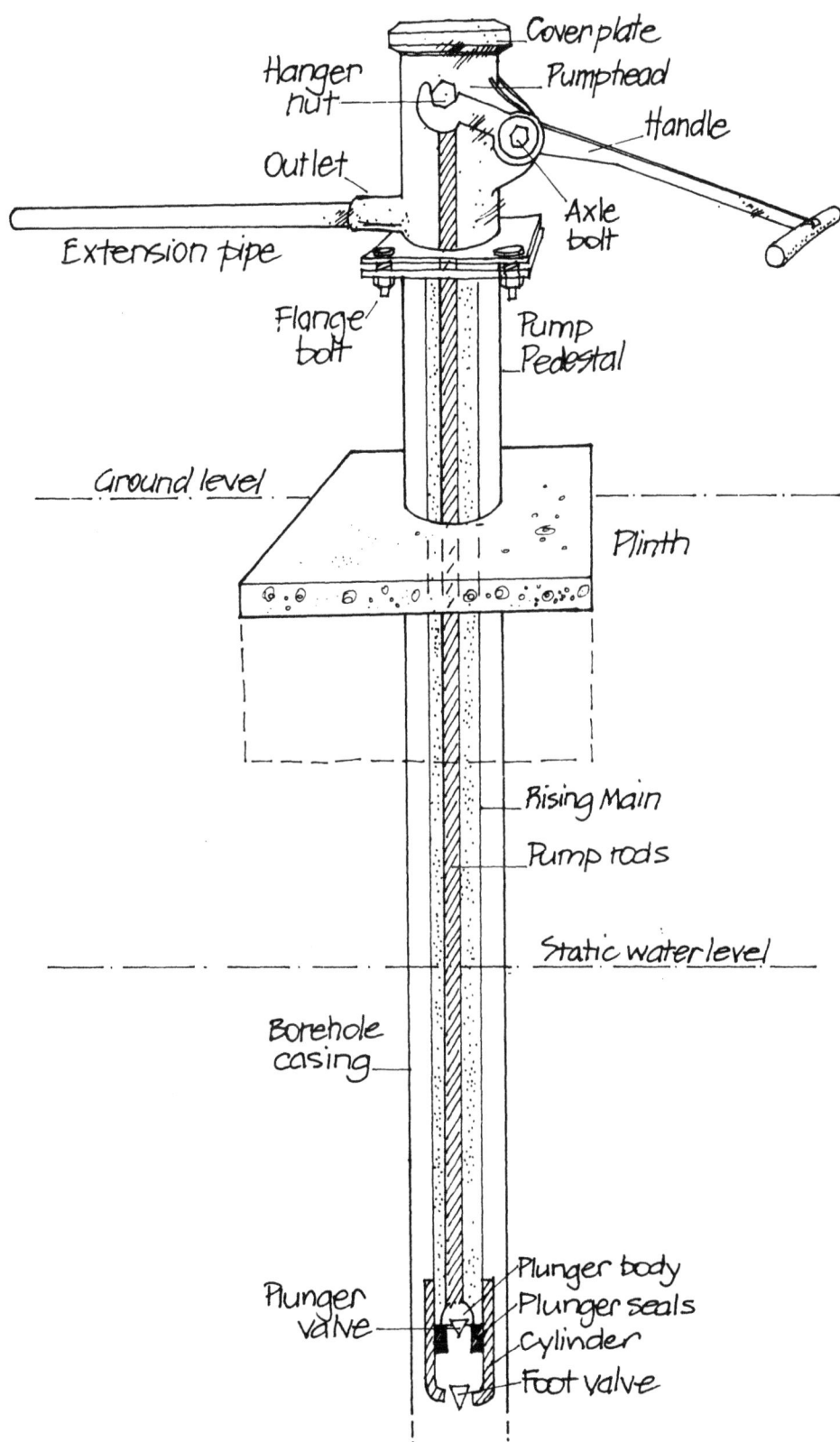

Figure 27a. MALDEV pump Ref (42)

Lifting the top cover provides easy access to the connecting rod (this is made up of various lengths screwed together to make up the required overall length). The assembled connecting rod is anchored to the pump handle using a bolted T-piece slotting into two recesses. Once again a sealed bearing allows for movement between the anchor bolt and the connecting rod. To remove the assembled connecting rod, the anchor bolt nuts are loosened and the rods pulled out by hand through the top of the pump. It is hoped to incorporate hinged rods in the final version so that no tools will be required to remove the connecting rod. In this way the possibiliity of community-based maintenance will be opened up.

It is intended that the down-hole components (piston and foot-valve) of the final version will be made from injection-moulded plastic. It is also expected that a foot-valve retrieving arrangement from the piston to the foot-valve will be developed so that, as the connecting rods are withdrawn, the valve is also removed.

The final objective is to provide a pump assembly such that a village pump caretaker, using perhaps only one spanner, will be able to withdraw worn or damaged down-hole components and quickly replace them with another cheap set of injection-moulded parts. Perhaps even Bata - the local shoe manufacturer - with its extensive retail network, might act as a distributor for the components. If the pump needs parts it will be up to the village caretaker to collect a few Kwacha (few US$) from the community and buy a new set of plastic components from the local store. In this respect Malawi appears to be leading the 'Village Level Operation and Maintenance' (VLOM) quest.

A key element in the pilot project's success has been the involvement of Malawian professional staff. All project management in the field is carried out by local staff with an ODA technical assistance team playing an advisory, backroom role. It is considered that the Malawian field staff are well able to deal with local leaders. The strategy of implementation with community participation includes giving each village the task of deciding where it wants the borehole or well to be sunk (subject to geological survey approval). It also has to select a village committee to take charge of well maintenance.

The project team estimate that the cost of a borehole, with pump, has now been reduced from $5,500 to $1,500, including a concrete apron and washing slab. (A dug well, constructed with community help, costs about $750).

The Malawi 'Integrated Project' approach for rural ground water supplies:

Malawi has (for the past fifteen years) been active in providing pure drinking water to rural areas through its gravity-fed piped water schemes. Their latest development programme in rural water supplies has focused on both low-cost boreholes and wells for obtaining drinking water from ground water supplies.

Problems with the previous provision of ground water by boreholes and dug wells included:
- (a) The borehole and dug well programmes had grown up independently without due consideration for the most appropriate abstraction method for a particular location (i.e. whether borehole or dug well).
- (b) The manufacture/construction and maintenance costs of boreholes and hand pumps had risen alarmingly (manufacture with construction cost: K5,000 to K6,000 hand pump; maintenance cost: about K200 per year).
- (c) The poorly designed boreholes were themselves a major contribution to the very high cost of hand pump maintenance.
- (d) The widely dispersed programmes resulted in inefficient use of vehicles - hence transport costs for construction and maintenance were high.
- (e) There was no maintenance structure for the dug well programme.
- (f) There was no involvement of the community at any stage of the borehole programme.

The proposal for an 'Integrated Project for Rural Ground Water Supplies' followed naturally. This approach involves the complete provision of ground water supplies in one area at a time by a project team, who would assess the current water provision by dug wells and boreholes and decide whether boreholes and/or dug wells were to be used for the project in any particular area. The intregrated project involved the sinking of low-cost boreholes using light-weight drilling rigs and a new borehole hand pump (the MALDEV pump).

The Malawi shallow-well pump

The Malawi shallow-well hand pump in used on the dug well programme. There have been various developments of the pump since the programme began in 1975 - from Mark I to a Mark VI version. A simplifying design feature of the pump is that there is no lever-type handle - it has a direct-drive T-handle. Since the pump normally operates at less than 6m pumping head, it does not require a piston seal. Even a badly worn piston works satisfactorily if operated at a high pumping speed.

Figure 28 shows the Mk V pump on a fully lined well. A Mk VI hand pump has been developed for higher pumping heads (6 to 10m). This pump uses a 45mm internal diameter cylinder instead of a 56mm diameter one, and it is necessary to have the piston below the static water level. The piston of this pump has a PTFE (Teflon) piston ring to reduce leakage.

Several features of the current shallow-well pumps require careful review. A common cause of failure is

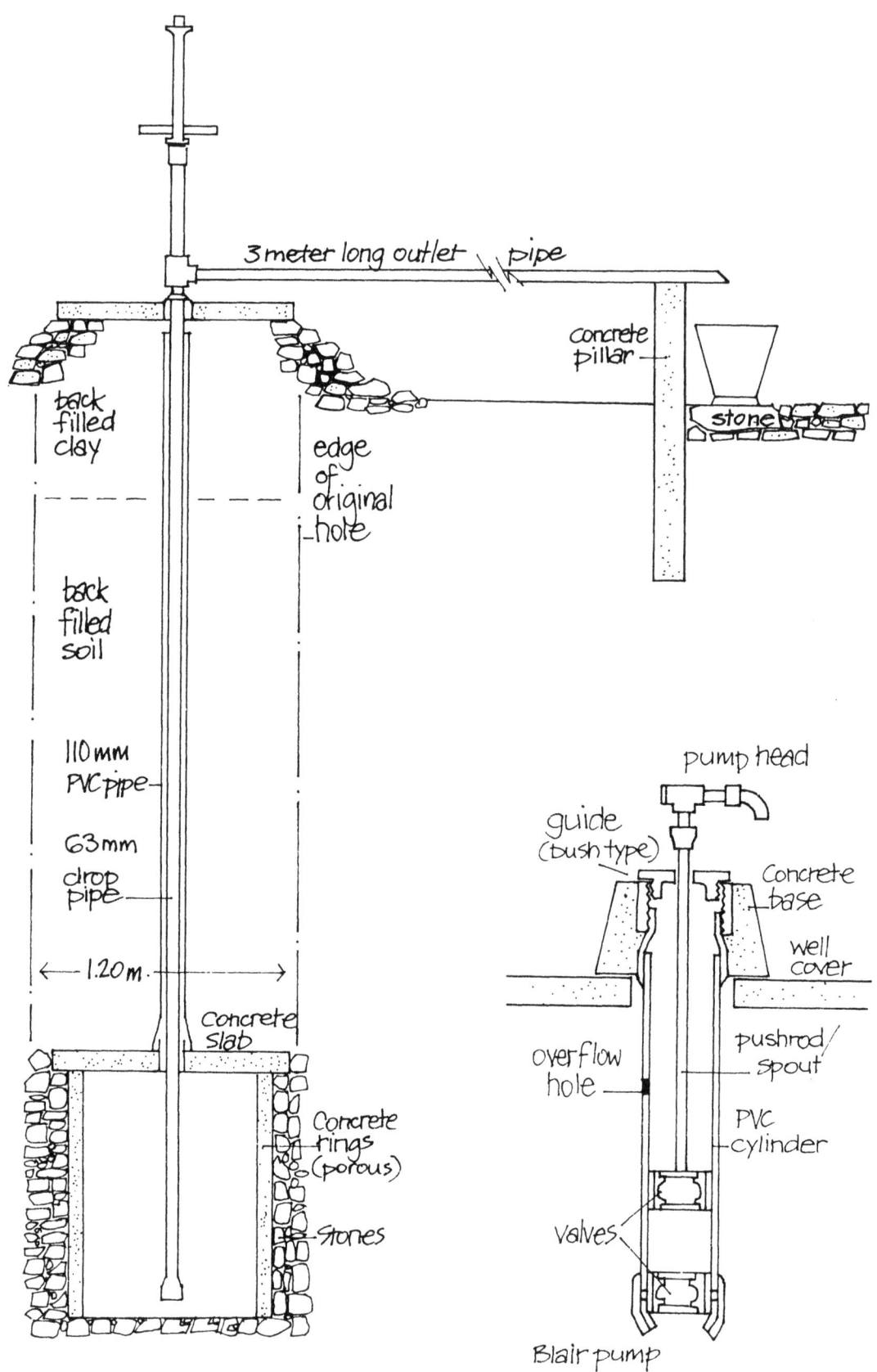

Figure 28. Malawi shallow well pump on fully lined well

breakage of the pump rod, most commonly at the PVC rod-steel handle connection (the original PVC 'T'-handle was abandoned due to frequent damage by abuse, and even being eaten by hyenas!) A related problem with the use of PVC rods is that replacement of broken rods requires the use of solvent cement. The problems of storing and using solvent cement probably rule it out for use by village-level repair teams.

(b) <u>The Blair pump (Zimbabwe)</u>*

This pump was designed by the Blair Research Laboratory, part of Zimbabwe's Ministry of Health (28). The original design was conceived as one requiring minimum maintenance and suitable for wells less than 6m deep. This meant that some components of the more standard (deeper well) pumps - a lever handle and water-tight seals - could be eliminated. These are two of the most troublesome components as regards level of maintenance and repair.

Figure 29 shows the original Blair pump, which was made from standard pipe fittings. This has now been almost exclusively superseded by a cheaper and more adaptable mass-produced pump.

A distinguishing feature of the Blair pump is its compact above-ground components. There is a galvanized iron 'walking stick' handle which doubles as the water spout. The pump only requires one person to dismantle.

Below ground, the pump consists basically of a stationary cylinder and a moving piston fixed to a hollow pushrod which is attached to the galvanized iron handle. On the up-stroke, water is drawn into the cylinder through the lower fixed foot-valve, this closes, the water is forced through the piston valve and up the hollow push-rod to the surface.

The mass-produced version is a 'minimum maintenance' design. A test-rig has been set up at the factory and after three months' non-stop pumping (six million strokes) the working parts have been claimed to show no detectable wear. Tests have also been made on pumps with heads of 6 to 11.5m using water to which soil has been added (1kg per 200 litres) - minimum wear was measured after 5.5 million strokes. It is accepted that field trials offer a more accurate assessment of durability. Some reports of the pump's operation in the field (from the Lutheran World Federation engineers in Zimbabwe) seem to suggest some maintenance problems.

*This is one of the pumps which features in the UNDP/World Bank field trials

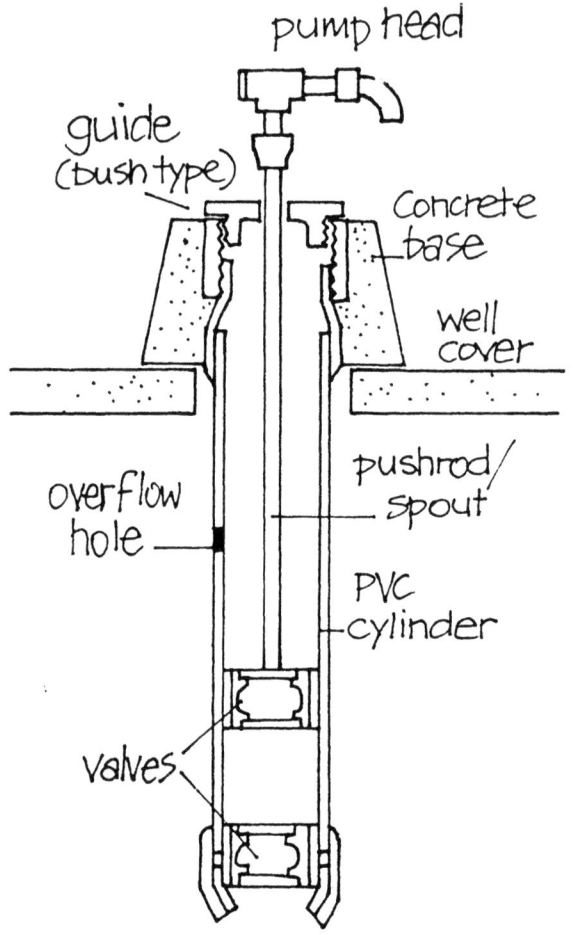

Figure 29. Blair pump (original version)

(c) The Rower pump

The concept of cheap, manually-operated tube-well pumps designed to irrigate small plots of land has become one of the focal points of irrigation development in Bangladesh in recent years. This emphasis has taken place in order to make water more accessible to small farmers, to make the poor rather than the wealthy receive the benefits, and to generate employment among the landless. From this scenario, the Mennonite Central Committee with the assistance of the Mirpur Agricultural Workshop and Training School directed its efforts at developing a pump envisaged by George Klassen, an MCC engineer (37).

The Rower pump (Figure 30) is a reciprocating-action piston pump whose PVC cylinder is inclined at an angle of $30°$ to the horizontal. A unique feature of this pump is that it is fitted with a surge chamber at the pump suction. This absorbs the impact of the accelerating and decelerating column of water within the tube-well pipe and so provides a steadier upward flow of water. This in turn enables the operator to make easier and quicker strokes. It has been

Figure 30.

found that a man can pump 50% more water in a given time using a pump with a surge chamber fitted. The ease of operation is remarkable - the addition of the surge chamber enables children (who would otherwise be too small to operate a conventional hand pump) to pump water quite easily.

Figure 20 shows the increased discharge obtained with the use of a surge chamber on both the Rower pump and the UNICEF New No. 6 pump (37, 31). (The Consumers' Association have also carried out laboratory tests on surge chambers for the World Bank.) Note that the use of a surge chamber could also be considered on other conventional hand pumps - especially to enhance the discharge for small-scale irrigation requirements.

The results of field tests carried out with the Rower pump by the Mennonite Central Committee indicate a prohibitive labour cost for the manual irrigation methods which rest or fall on the question of labour costs. Where labour costs have to be taken into account (and where these are relatively high) then the case for manual irrigation becomes less favourable.

Rower pump developments

In 1981, the Mennonite Central Committee (MCC) and the Mirpur Agricultural Workshop and Training School (MAWTS) conducted an experiment comparing three manually-operated irrigation pumps: the 2" Rower pump, the treadle pump manufactured and marketed by Ranpur-Dinajpur Rehabilitation Services (RDRS) and the new No.6 Bangladesh pump. It was found that the Rower pump continued to operate usefully at lower static water levels (down to 8.5m) than the other pumps tested. Naturally in irrigation pumping situations the tubewell requires to have aquifer conditions which do not significantly restrict water inflow to the well.

Since 1980 the Mennonite Central Committee have established a marketing effort geared to the sale and distribution of the Rower pump. For the 1980-81 season total sales were 415 pumps. During the 1981-82 season 650 pumps were sold and installed in three districts, with a further 150 pumps distributed by MAWTS in three other districts. The sales target for the 1982-83 season was 1000 pumps.

The MCC consider that their experience in introducing the Rower pump for tubewell irrigation constitutes a test of the principles generally involved in the introduction of new technology. The MCC initiative in irrigation development involved a package of three new constituents - the Rower pump, PVC tubewell and filter. This was a situation which required special efforts since without proper field training and installation there would be a danger of rejection of the whole package even if only one constituent failed.

Training was an important aspect in introducing the new technology. It was essential to provide good examples and 'hands-on' experience during the training period. Field testing is also essential for product development. MCC did extensive field testing, while maintaining close communication with the users. This led to improvements and simplifications being made to the original design.

When the Rower pump was first introduced it was installed above ground. However, recently more and more Rower pumps have been partly buried in the ground with only the discharge pipe exposed. This has the advantages of keeping the lift to the absolute minimum, eliminating the danger of knocking or damaging the surge chamber, and reducing the risk of theft or vandalism.

The Rower pump is vulnerable to damage if the pump handle strikes the end of the pipe. The latest design consists of a flanged steel insert to eliminate damage of this kind.

The Rower pump is now also manufactured by a UK company (SWS Filtration Ltd.) Two operating options are suggested: a standing position or a sitting one. The standing position is recommended for household (domestic supply) operation. The sitting position is recommended if the pump is to be used for irrigation. (As mentioned already the pump may be completely buried in the soil, leaving only the outlet above ground level.)

The SWS Rower pump has a maximum effective suction lift of up to 3.7m. Within this range an output of up to 1.5 litres per second is claimed (achieved by an 'average healthy man'). This is a 'water watts' output of 55W which suggests a manual input which could not be sustained for lengthy periods. Women and children would be expected to pump less water, but even a child pumping slowly can operate the pump and produce more water than that obtained from most other hand pumps.

(d) The Tara pump: the new Bangladesh deep-set hand pump - the most recent UNICEF development (20).

During the 1970s there was a concentrated effort made in Bangladesh to expand shallow hand pump tube-wells. The goal was to deliver some 50,000 units annually by 1979. These tube-wells used a 'suction-type' hand pump - the Bangladesh New No.6 Pump. This was developed taking into account local manufacture, distribution, installation, operation and maintenance problems associated with previous pumps. As a result of increasing pressure by Government to attain self-sufficiency in agricultural production there is an increasing use of motor-driven, deep-well pumps. These will eventually lower the water table below 7m (the limit for suction-mode pumps used in the shallow-well programme). Hence the drying out of shallow tube-wells greatly increases the area requiring 'deep-set' (lift/force type) hand pumps. It is estimated that up to 50% of the area will be affected in this way, requiring over 700,000 'deep-set' pumps in those areas. The existing pumps of this type are expensive - a typical one costs some US$2,000 when one considers the real cost of maintenance in the first two years, since they have a very poor maintenance record. They are difficult to sink quickly and caretakers cannot be involved in their maintenance. What is needed is an inexpensive pump (about US$300) which can be quickly installed, and which requires minimum maintenance, the latter to be carried out by village caretakers. (It should also be produced in Bangladesh and be capable of operation by children).

The India Mark II deep-well pump has proved to be very successful for areas where a 50m lift is required and where a maintenance back-up system can support it. Neither of these exist in Bangladesh where the maximum lift presently stands at less than 15m and the maintenance service has so far proved inadequate to cope with the few deep-set hand pumps that are already in operation.

The Tara Pump (Figure 31) - previously the Bangladesh deep-set hand pump - has evolved from experience with PVC pumps in Malawi and Ethiopia.

The first working prototype has been installed. It is direct drive (i.e. no lever) thus eliminating the wear and need for lubrication of rotating parts. All 'above ground' components are made from locally-fabricated galvanized iron, to avoid degradation of PVC in the Bangladesh climate. The piston and foot-valve are those already used in the Rower pump. The well can be installed as quickly as the shallow tube wells using the sludger method, and it can be maintained entirely by village caretakers.

(e) <u>The Rangpur-Dihajpur Rehabilitation Services (RDRS) treadle pump (with bamboo tube-wells)</u>

Since 1971 the Rangpur-Dinajpur Rehabilitation Services (RDRS) has been involved in assisting small-scale farmers (since 1981 the target group was farmers with 2 acres or less - before 1981 the 'ceiling' was 3 acres). The government's agricultural extension programme was supported through research and training, appropriate innovations (pumps with bamboo tubewells for irrigation, and other farm implements) and the introduction of new crops, e.g. wheat.

RDRS has worked on simple, manual and affordable agricultural implements, one of the most obvious successes being the bamboo treadle pump. Many models were tested under field conditions and a production model - a twin-cylinder pump - was available in November 1980. During the next two years, nearly 7000 of these irrigation pumps were installed in the field. The simplest twin cylinder model delivers $5.5m^3/h$ from 5.5m. Note that this data gives a 'water watts' output of approximately 80, which suggests that the human power input is considerable - perhaps about 140 watts. This level of human effort could not be sustained for extended periods even with the weight-shift treadle-action operation of this pump.

The metal and plastic parts of the treadle pump cost only US$7. The farmer provides the bamboo and hires a tubewell sinker who charges an average of US$0.30 per metre for the installation. The total cost (pump plus typical installation) is about US$13-17. In comparison with this a typical commercially manufactured hand pump costs about US$100 and provides a flow rate of about $2.7 m^3/h$ only, at a pumping head of 5.5m.

To ensure optimum results, farmers as well as local sinking groups are trained to install the pump properly and to treat the bamboo with chemical preservatives. Small farmers purchase their own pumps and pay for the installation (they benefit from a small subsidy of US$3).

At present, only 4.35% and 3% of the arable land in Rangpur and Dinajpur districts, respectively, is under irrigation farming. Hence there is considerable scope for increasing food production by enabling the farmers to grow two or more crops per year. During 1982, the programme sold 4733 irrigation pumps (plus 869 drinking water pumps). To meet the growing demand for pumps, in addition to the main RDRS workshop in Rangpur, three other private workshops have been established with RDRS assistance. The four workshops have a production capacity of 3000 pump heads per month. There are seven models of the irrigation pump, and one domestic model available.

Some comparison between RDRS pumps and diesel pumping is relevant here. The subsidized cost of a diesel-operated pump (0.75 cu.sec) is US$1040 - 1250, while 10 treadle pumps with bamboo tubewells cost only about US$125. The field life of a diesel pump is about 3 to 5 years, while the bamboo tubewell (with the bamboo treated with preservatives) is expected to last over 10 years (untreated bamboo lasts only about 4 to 5 years). These reasons plus the rising fuel costs make RDRS pumps more appropriate and viable for the majority of small farmers.

WE WOULD LIKE TO OBTAIN FURTHER INFORMATION ON SOME OF THE PUMPS INCLUDED IN THIS CHAPTER - PARTICULARLY DETAILS OF RECENT R AND D FIELD DEVELOPMENT. SEE APPENDIX 1.

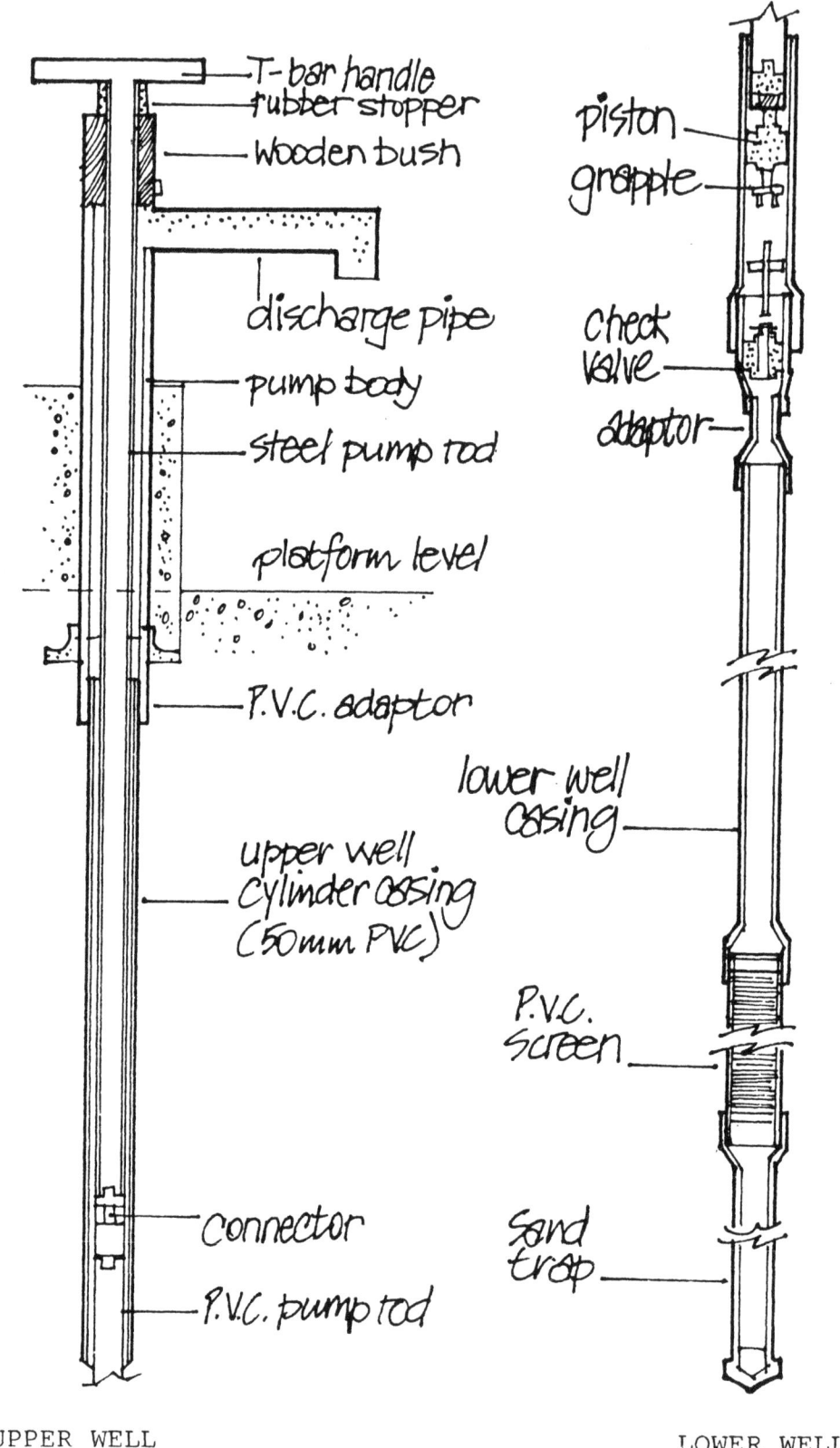

Figure 31. Tara pump (Bangladesh deep-set handpump Mark I) Ref. UNICEF, Dhaka.

TARA Pump update

Forty TARA pumps were installed in stages during 1983, and fifty were sent to UNICEF in Sri Lanka for field testing.

Relatively minor changes were made to several components in response to problems which appeared during the field testing of 40 prototypes. Modifications to the original design include a larger diameter (2 inches) discharge spout, a larger diameter PVC pump rod, more robust pump rod connectors, a redesigned check valve, a PVC top guide bushing lined with an oil-soaked wooden sleeve, a shortened T-bar handle, and higher, thicker-walled leather cup seals.

Six tubewells failed as a result of choosing a silty aquifer, and 'third generation' TARA pumps were reinstalled on new tubewells. One of twelve first generation pumps is operating after a year of heavy use. The remainder were retrofitted with second and third generation components.

For the TARA pump to be applicable for rural water supply in Bangladesh on a mass scale it must withstand years of heavy use without increasing demands on the already over-extended public sector. The question which follows is whether the users are able routinely to change valves and seals.

The PVC (pipe) pump rod is undoubtedly the most vulnerable element in the system. PVC is not an engineering polymer specified for use as a tension member subjected to repeated stress reversals. Fatigue failure is therefore a distinct possibility. The question is: 'after how many stress reversals and at what peak load?' The planned three years of field testing, accompanied by accelerated rlaboratory fatigue testing should provide definite answers. Meanwhile, alternative polymers are being investigated.

Buckling of the pump rod is apparent on the down-stoke, and abrasion marks appear on the pump rod. No doubt abrasion also occurs on the rising main. Eventual perforation of the pump rod or weakening to the point of failure is possible. Guides on the pump rod are being considered to minimize this buckling.

The UNICEF Regional Project Officer in Bangladesh concludes that a direct-drive low-lift hand pump is a viable concept with distinct advantages over conventional hand pump technology specified for heads up to 20m - major problems with the system do not appear likely, except in the case of unknown fatigue resistance of the PVC pump rod. If PVC proves to be unacceptable, appropriate material substitution could rectify the problem, although at a somewhat increased cost.

What is 'Low Lift'?
'Low lift' pumping generally evokes an image of a suction pump. Indeed 'suction pump' and 'shallow-well pump' are synonymous terms. With suction pumps the piston is located

above the static water level and the lift is limited to about 6 or 7m. However, the same pumping elements located below the water level could lift water with the constraint, for all practical purposes, of the structural/material limits of the pumping elements and the amount of power available to move the piston upwards. This means that with a judicious combination of cylinder diameter, stroke length and a light-weight pump rod, a simple direct-drive piston hand pump of adequate capacity can be designed to operate at a head of 20 metres or more. This is the basic concept behind the Bangladesh TARA pump (20).

2.8.2 A possible R & D project on human-powered irrigation pumps

A field assessment of projects which have involved the use/development of human-powered irrigation pumps is a necessary pre-requisite to development work in this area. This assessment should include the following types of pumps:

- chain and washer pump (comparing the performance/life of any 'low-cost' versions which have had adequate field testing with, say, factory-produced versions such as those produced in China. (See references 8 and 9 for details of the hand-turning, two-person design.)
- diaphragm element pumps - further information on field trials carried out on diaphragm pumps, such as the AIT-developed version of the IRRI design, and the version from the Bangladesh Rice Research Institute (also based on the IRRI design).
- treadle mechanism pumps - 'low-cost' pumps developed at the Rangpur-Dinajpur Rehabilitation Services (RDRS), Bangladesh (by Gunnar Barnes).
- the Rower pump - a follow-up to the field trials and project implementation in Bangladesh by the Mennonite Central Committee, via the Mirpur Agricultural Workshop and Training School (George Klassen and others). This work includes economic studies but recent data is not to hand. (N.B. the Rower pump featured in comparatively recent laboratory tests by the Consumers' Association. The Consumers' Association has also carried out tests on surge chambers - an integral component of the initial Rower design.)

Once the above pumps have been fully assessed in the field then it should be possible to identify possible future developments. It may be necessary to carry out further controlled field (and laboratory) tests. Since irrigation pumps will probably be subjected to fairly sustained pumping the ergonomic aspects of design will be important. In this respect the above pumps may prove to be ineffective. There may be scope here for novel designs, e.g. foot operation.

3. HUMAN-POWERED WATER-LIFTERS: SOCIAL ASPECTS

3.1 Introduction

This section is concerned with the social aspects of water-lifting developments - particularly related to programmes involving community participation. While there is a significant volume of literature on community participation in the sphere of water supply and sanitation, few writers refer specifically to water-lifting devices. Nevertheless, it is principally through the 'community participation' approach that projects such as those introducing hand pumps are now executed.

3.2 Community water supply systems

There is considerable evidence which points to the need for radical changes in the way in which decisions are made and projects implemented within the general field of community water supply systems. The easiest and most obvious way for development agencies to respond quickly to the demands of the present UN Drinking Water Supply and Sanitation Decade, is to put money into hardware - dams, pumps, pipes etc. There is often not sufficient attention paid to the software side of programmes - planning, management and organizational aspects. The very urgency with which a typical water programme is often pursued automatically works against its own chances of success by reinforcing administrative structures and procedures which are singularly unsuited to the promotion of the declared development objectives. Such programmes divert resources away from the often tedious and time-consuming tasks of pre-project community consultation, participative planning, improving local management capabilities and performance through field training and procedural/organizational reforms. The irony is that, on paper at least, community participation has been emphasized as a major factor in successful rural water development programmes. Community participation, with its related ideas of 'user-choice' and 'bottom-up' approach is one of the principal concepts being stressed during the current UN Water/Sanitation Decade.

There is an immense amount of literature on the subject of community participation and we shall illustrate the general themes and issues which surround the approach with case-study experiences.

Although Isely (19) refers to 'Community Organisations', the model which he proposes is essentially a 'community participation' one. Isely takes the definition of community organization to be a process intrinsic to a population in which existing social structures mobilize for action in the face of defined problems. The process does not involve the creation of new structures, nor the replacement of existing ones, but rather the enhancement of existing ones.

Isely stresses that community organization means working within the framework of resources and constraints in the community which vary greatly from one area to another, and workers must take account of these various requirements.

Early steps in helping community organizations include meeting community leaders, at which point the crucial process of building confidence begins, gathering information and identifying existing structures which can serve as vehicles of community development. Once the group is organized the worker will help it to agree on a set of problems of greatest importance to the population. One of these problems will then be selected to be dealt with first. Usually the one chosen will be the most pressing to the dominant faction, or the most amenable to solution in the local situation. The worker and others must then inform the population about the problem and raise their level of knowledge to it and its ramifications. Pacey (32) and Holmberg, (14), amongst others, also advocate the use of existing and potential communication channels in the adoption of innovations and new behaviour, for example local singers and story tellers, drama groups, folk operas and posters.

Local resources will be assessed to determine the extent to which money, labour, materials and technical expertise can be obtained locally and the solution to the problem is divided into several stages, each being given a concrete objective, the time limits and necessary labour and materials. The responsibility for executing these stages is allocated to individuals and groups and after implementation, achievements are matched against objectives in the evaluative stage.

Trained community workers are seen as essential to successful community organization and they require diagnostic, supportive and evaluative skills. They must be able quickly and easily to elicit information about the community which is relevant to the proposed action, and establish trust and help group discussion while allowing community leadership to emerge. They must also stimulate the community to make evaluations, thereby strengthening commitment on their part.

Another essential factor promoted by Isely (19) is that agriculture, education, public works and health must all aim to facilitate community organization rather than achieve independent technical objectives. It is claimed that with the success of each development stage or objective, the capacity of the community to absorb new technical inputs and face new adaptive challenges increases.

The results of community diagnosis should, according to Isely's model, indicate which of the available technologies are economically affordable, socially acceptable and technically maintainable for water supply and sanitation. Isely argues that there is an urgent need for manuals with non-technical drawings and a minimum of written explanation for community members themselves.

Although this model has developed out of observations made of a wide range of case-studies in the field of water supply and sanitation, a number of the assumptions are highly contestable. The strong emphasis on time schedules and speediness of execution is refuted by evidence that projects where community participation has achieved success have not put any emphasis on time and speed of execution.

Feacham(11) argues that concepts such as 'community participation', 'user choice', and 'village level' are used in an over-simplified manner and direct attention away from the fundamental political and administrative realities that primarily determine the success or failure of rural water and sanitation programmes. The notion of a small governmental input into a rural community catalysing a chain of self-sustaining development activities on the part of a village has been discredited in his view. The currently fashionable and more elaborate forms of community participation require a cadre of staff and a sustained interaction between government and community that is totally beyond the means of most governments to implement.

Feacham and other critics throw doubt on the relevance of this approach, pointing out that community participation has never been a major component of water supply and sanitation development in the industrialized West.

The community participation model is also criticized on the grounds of cost, because while savings may or may not be made in costs of construction, operation and maintenance for a particular project, the overall programme costs of the government will always have to be increased by the remployment, training, supervision and transportation of the teams of community-level workers that are necessary for successful community involvement.

Opinion varies as to whether the relationship between government and community should be standardized or flexible. Feacham's view is that while it may be desirable to maintain flexible village-government and intra-community relations it may be impossible to do so on grounds of cost and practicability and could be administratively infeasible. Given that most water and sanitation agencies within developing countries are striving for a high degree of standardization as a partial answer to their managerial, administrative and financial problems, they may well be right to resist any attempt to encourage a flexible approach to participation on a community-by-community basis.

Community participation involves joint action with some external body, commonly the government or one of its manifestations, and the willingness of communities to develop links in this way is partly dependent on the political context and the history of previous relations with the government. In many cases, due to strong opposition to the ideology and activities of a government the people, far from wanting to participate in joint development projects, have strong motives for attempting to sabotage such projects.

Whyte and Burton (47) examine the problem of maintaining efficient water supply systems, and note that of the new systems built in the last twenty years about half are not working at all and another 30 per cent are working only intermittently or inefficiently. Instead of looking at problems concerned with the 'delivery system' they look at the product from the consumer's point of view and ask, 'What are the consumer's needs and aspirations in relation to water?' This is known as the 'User Choice Approach'. Whyte and Burton look at the processes involved in community decision-making and illustrate that decision-making styles vary culturally and geographically, often showing considerable variations within a local area.

In their view self help schemes are still conceived and implemented within a 'delivery philosophy' with the community being given choices only in details rather than fundamentals. Traditional community decision-making styles must be given due consideration if a project is to be successful. Holmberg attributes the failure of a wells project in Peru to the total lack of consultation or involvement of village leaders. The local council appointed to organize community labour had no native-born members and no-one to represent the leadership of the village. Further, they omitted to consult a certain local farmer whose prestige was high in the community and who knew much about well-digging and local water conditions. This created hostility towards the programme and the types of decision that were made. Scotney found that where expectations of secondary users of a new water supply are frustrated negative attitudes towards maintenance can follow. Prohibition of irrigation and cattle watering in supply schemes in Kenya was a factor which contributed to damage and non-payment of water rates.

In emphasizing their view that ideas and innovations will only diffuse through the community successfully if they are introduced and promoted through traditional community decision-making mechanisms, Whyte and Burton do not address the problem that decisions reached in this manner may not benefit all groups as regards access to improved water supplies and reduction in drudgery. The notable case here is the viewpoint of women as carriers and suppliers of water which is often overlooked. Dube's study (10) gives weight to this point. He found that the value system in the villages was an 'undermining influence' in that no education could be given to women or the lower strata which meant greatly reduced project benefits to these groups.

An analysis of the social role of water in traditional rural communities is incorporated into Whyte and Burton's study with reference to Wittfogel's 'Original Despotism' and his theory of water control as the basis for early civilizations. Wittfogel focuses his attention on water for irrigation, but water for domestic purposes acts as a force for social integration and social differentiation as well. The water source is often a meeting-place for the community and communal ownership of the source may articulate the boundary between the in-group and the out-group. Therefore when a new water supply system is introduced it may bring with it not only improved facilities but also a new social order. The change commonly affects the balance of power away from traditional leaders towards the literate and politically sophisticated and the external relations of the community are altered, usually by increasing the dependency orn national and regional government and decreasing the ability of the community to act independently in relation to others.

The failure of villages to maintain their new water systems can frequently be diagnosed as an adherence to their pattern of social relationships to which the new system was insensitive and disruptive.

Cohen and Uphoff (7) attempt to clarify issues such as how 'participative change' projects are distinguished from those which are 'technocratic' and 'centralized'. The three key dimensions are, in their view, firstly the subject matter of participation whether it is decision-making, implementation or evaluation and secondly who does the participating. It is emphasized that the rural poor do not see themselves as a homogeneous group, and usually clear distinctions may be made between various groups. The third dimension is concerned with how the process of participation takes place and tries to distinguish between forms of participation which are 'initiated from below, voluntary, broad in scope.....' from those which are initiated from above, and coerced.

It is suggested that modes of participation, as defined by these three dimensions will vary with different projects and their environments. Cohen and Uphoff conclude that it is essential to consider the purposes of participation because the authorities may have a very different purpose to that of the community concerned. They warn of the danger of participation becoming a fad.

Agarwal (1) looks at developments in this field in Malawi. This is one of the countries most certain to meet the UN's 'Clean Water For All' target for its rural population by 1990. The country has only to step up the current water supply programme by 13 per cent a year, a remarkable achievement particularly because 90 per cent of Malawi's 5.2 million population lives in the rural areas. The key to success has been community participation. A project is initiated when a request from the District Development Committee consisting of MPs, local Party leaders and traditional leaders, is received by the Ministry of Community Development and Social Welfare. The Ministry draws up plans and village leaders establish a committee to organize construction which is shared among the villages taking part in the scheme. The site for each village tap is chosen by the villagers, who construct a concrete apron around it and a concrete drain into a soakaway. At the inauguration ceremony the villagers are instructed in how to operate the tap and advised of the costs they will have to bear if it is broken. These gravity schemes are maintained by village volunteers who are given a two week training programme funded by the government.

Bharier argues that the motivation of villagers in Malawi to carry out self-help schemes is clearly linked to their recognition of the health benefits obtained as a result. The success of the schemes can also be partly attributed to the fact that, in comparison with other African states, Malawi has had a long period of political stability as a consequence of its one party political system and Dr. Banda's long presidency.

In Tanzania a two-tier type committee system operates with Divisional Water Supply Boards which are elected from councillors, political leaders and village officials and these administer the allocation of materials, with the power of withdrawal in the case of substandard village performance. Representative local committees made up of household representatives and political leaders are responsible for the organization and supervision of community participation. Nevertheless, whilst the programmes were administered from a 'grass roots' level, Matango (25) in his report on schemes in the Lushoto District, found that there were other constraints which operated to mitigate against the success of projects. There was a lack of planning as regards potential water uses and volumes required and - a problem which is frequently emerging in community participation programmes - a shortage of labour. In Tanzania this shortage was exacerbated by prolonged religious fasting which affected the stamina of the workforce who were already spending two days per week working on other community projects. BURGEAP (3) report similar difficulties in other parts of Africa, namely Niger, Upper Volta and Tchad. They found that the workforce required for community participation is different to 'normal' village labour and that the traditional division of labour makes water supply contributions less interesting for males. This provides evidence against the view of community participation as a process which strengthens existing social structures.

BURGEAP's study of community participation schemes in Togo, Tchad, Upper Volta and Niger reported mixed results and suggests a number of general conditions which are deemed to be necessary for effective involvement of the population. It found that only in Niger and Upper Volta was the experience successful and, like Isely, they stress that it is essential to have efficient and dynamic co-operation between agencies and groups involved in a project. The presence of an urgently-felt need by the villagers is also crucial especially in the case of water quantity improvement. Projects are generally more successful with sedentary rather than nomadic groups probably because the latter tend to have strong social structures, and significantly fewer contacts with rural extension. The study concludes that nomadic groups have stronger social structures and are therefore more robust and able to solve their own problems. Community participation is used when a group's indigenous problem-solving mechanism has broken down or where the group is unable to adapt sufficiently to deal with the changes required.

Khare (22) and Dube (10) both provided case studies of projects in India where innovation has resulted in increases in inequality. These studies, while not referring to water-lifting devices specifically, illustrate a recurrent theme in the community participation literature. Khare found that religious and caste attitudes played an inhibitive role in the lack of adoption of introduced innovations; yet, he concludes, where such attitudes are 'passive', innovations are often co-opted and manipulated in existing intra-community conflicts and competition for village leadership. The WHO literature on community participation suggests that in such cases material inputs should be suspended until 'organizational' problems have been solved.

Dube's study highlights the fact that innovations may lack specific cultural, economic, educational or sociability functions. In the case of new sanitary wells the innovation was culturally acceptable but given that the new wells required more energy for operation and meant longer waiting times, the upper castes were able to assume a supervisory role and exerted pressure on the lower castes to perform extra physical work without compensation. The upper castes were thus able to exploit the situation to their further advantage.

Khare (22), Dube (10), BURGEAP (3), Fearham (11) and Rogers (35) all raise the issue of community participation programmes being characterized by a mistrust of government intentions. Whether individuals or committees will wish to participate with government depends in part on the history of their previous relations. If there is strong opposition to the activities and the ideology of the government, attempts may be made to sabotage projects.

Rogers (35) looks at the perceptions of the community of the national and local media and finds that where there has been a high degree of government control a negative influence on the perceived trustworthiness and competence will result, which reduces their possible impact in diffusing information about new technology and innovations.

3.3 Water supply for irrigation

Water-lifting devices are not limited to use in drinking water supply systems but are also important components of irrigation schemes. It is therefore relevant to look at the social impact of irrigation development.

There is a growing awareness that much of the poor performance of irrigation schemes stems from fundamental weaknesses in planning and management which no amount of investment in technological hardware will overcome on its own. There is a need for radical changes in the way decisions are made in irrigation planning and the provision of domestic water supplies. The easiest and most obvious way for development agencies to respond quickly to the demands for new investment is to put their money into hardware such as dams, pumps and pipes. Little attention is paid to the 'software' side. For example, the management of water distribution for irrigation is fraught with problems, since water in scarce supply inevitably becomes a focus of competition and potential conflict among its users.

Invariably the very urgency with which a typical water programme is pursued works against its own chances of success. It strengthens the already over-dominant role of design and construction engineers in the decision-making process, at the expense of others such as local government organizations, and agricultural staff and farmers. It diverts resources away from the often tedious and time-consuming tasks of improving management skills through field training and reforming procedures and organizations. By encouraging heavy reliance on central government and donor agency funding, it further inhibits the scope for local participation in planning construction, thereby compounding the difficulties of promoting communal responsibility during the subsequent stages of operation and maintenance.

An increase in local participation in planning and day-to-day management and the creation of support organizations, which can respond to needs while exercising sufficiently impartial authority to minimize inter-community conflict, is needed to improve the success rate of water development schemes. Development agencies are only just beginning to accept that major changes need to be made to their policies and practices.

Chambers (5) makes the point that attention tends to focus on the hydrological, engineering, agricultural and economic aspects of irrigation and the relationship between people and irrigation is usually ignored. Water is a scarce resource over which people compete and the benefits it brings should be optimized in relation to other resources. In the dry zone of Sri Lanka there is evidence that water is more limiting than land, although shortages of draught power and labour are other constraints. In parts of North Arcot, India the scarcity of water is even clearer and more acute and the ground water level is dropping rapidly because surface irrigation water from tanks is not adequate for a second crop, and the numbers of wells and pumps being installed are increasing dramatically.

A central and universal issue in the distribution of irrigation water is who gets what, when and where. The allocation and appropriation of water has two stages: the first involves decisions regarding areas of operation and timing; the second, the actual allocations and appropriations. These decisions raise acute and complex questions of equity which are interlinked with patterns of wealth and power within irrigation communities.

The developing world is rich with a variety of local organizational arrangements for managing irrigation water. These systems have been constructed, maintained and operated by local people, occasionally with outside assistance of some kind. Studies of these systems often fail to include critical information regarding basic elements of the situation which would give a more complete picture of the environment in which the irrigation organization operates. Nevertheless it is possible to see in these various cases some common themes and principles that are relevant to current efforts in irrigation development.

Several countries have extensive portions of their irrigated land served by traditional systems and are interested in improving the facilities and operation of these systems already in place. Other countries are investing considerable sums in the development of new, small-scale irrigation systems. The traditional irrigation systems of South East Asia are particularly interesting.

Coward (9) describes how farmers have designed, operated and maintained their own schemes using simple diversion weirs. These schemes are designed for between 200 to 300 hectares of land. Coward selects three crucial features of these community irrigation schemes. The first is that the leadership is accountable. Leaders are selected by the group, their performance is reviewed by them and they are compensated for their services directly by the group. Secondly, the systems, while on a small scale, are divided up into sub-units each with its own leader, and 'management intensity' is very high. The third feature is that the boundaries of the systems rarely coincide with village boundaries and are channel-based instead. The equitable distribution of water and the control and resolution of conflict has evolved over a long period of time and has been very successful until governments have tried to incorporate communal schemes into larger projects. The indigenous groupings and mechanisms for control have then started to break down.

The Philippines government has an active programme of technical and financial assistance to communal systems throughout the country. In many cases this assistance will provide significant improvement to critical elements of the physical infrastructure: the replacement of a bamboo diversion structure, or the installation of a permanent headgate to reduce the risk of flooding damage to the canal system. This represents a strategy in which traditional systems are further mobilized to contribute to national goals.

The introduction of improved, sophisticated irrigation structures into traditional systems may serve to make them more dependent upon the outside bureaucracy. In contrast to the bamboo and stone weir of the traditional systems, the concrete weir may require financial resources, masonry skills and other pre-requisites not available within the local community.

In the Philippines context material assistance is being given on a loan basis. The government is naturally concerned with ensuring the repayment of this investment.

Thus, each indigenous system receiving government assistance is required to organize formally in terms of officers required, committees to be organized and other procedural rules. The need is to implement and provide assistance in as unobtrusive a manner as possible.

The current indigenous models of irrigation organization that exist throughout the developing world represent an important pool of experience and examples for the current efforts at irrigation development. Clearly, these indigenous models are not equally applicable to all irrigation projects but they demonstrate locally-derived solutions to recurring problems in our modern systems. It may be possible to incorporate some of the contextual properties characteristic of indigenous systems into modern irrigation schemes. This is certainly the case for small-scale 'modern' systems.

Planners and policy makers may have some erroneous ideas about indigenous systems, if they are aware of them at all. These myths can be dispelled only through the accumulation of more detailed information about a wide array of local systems operating in various natural and socio-economic conditions.

4. ANIMAL WATER-LIFTERS: TECHNICAL ASPECTS
4.1 Animal power: what power is available?

What effective sustained power output can be expected from various draught animals? This is a rather difficult question to answer since there is little reliable data available which is relevant to a given animal in a particular condition, and environment. There is 'rule-of-thumb' data based on animal weight; there is field data which is most unreliable/incomplete - much of it does not indicate whether the power output attained is sustainable for a lengthy or for a limited period; and there are some controlled laboratory tests which invariably give data based on animals in prime condition, with proper feeding and working in 'ideal' environments.

There is certainly a need for reliable data - not only laboratory tests but more importantly field tests on draught animals in developing countries (in whatever condition they are - provided this can be specified in some way).

The variables involved are considerable - the weight, age, condition, feeding (quantity/quality), the nature of the work (continuous and sustained/short-term), the environmental conditions.

After a detailed study of the literature related to animal power the author has concluded that the data in Table 5 seems to offer a reasonable assessment of animal power. It is by no means 'high certainty' data - it is probably the best that is available and represents a fair degree of agreement of (fairly reliable) data from different sources.

Table 5. Animal Power

Weight of Animal	Single Animal	Pair of Animals*
kg	W	W
325	190	340
400	235	425
500	295	525

* The effectiveness of two animals harnessed together is less than what would be expected from the two animals working separately (an overall 90% effectiveness is assumed).

Figure 32 shows the 'ideal' (100% transmission and mechanical efficiencies) Q-H (flow rate-head) characteristics for a single animal giving 200W and 250W power output. This Figure also shows the performance of actual animal-powered devices - these will be considered in the next section.

4.2 Review of traditional animal-powered water-lifting devices

The water-lifting devices which operate by means of animal power generally fall into two categories: those involving linear movement based on the use of a pulley, rope and container, and those involving rotational movement utilizing a rotating wheel or drum. In the first category the animal moves forward and then back and there is invariably intermittent operation. In the second category the animal moves in a circular path with the potential for continuous water-lifting.

4.2.1 Rope and pulley arrangements

The 'rope and bucket' lift (Figure 33) is a device which is much in use in parts of India and throughout the near East. It is much used in those areas where the water table is lower or near the maximum practicable lift of the Persian wheel (about 9m). The device (referred to in India as a 'mohte') consists of a bucket or bag (originally made of leather but metal buckets have now gained favour) attached to a strong rope which passes over a pulley set in a framework and positioned over a well. Usually a pair of bullocks are hitched to the arrangement. The animals walk down a slightly sloping path away from the well while pulling up the full bag. The effort required during the lift is variable: while the bag is submerged the force required on the rope is fairly small; at the point at which it clears the water surface considerable effort is needed, which involves the attendant assisting the animal(s) with a likely instantaneous power output in the region of 400 to 500 watts per animal. Two operators are needed for this kind of water-lifter - one to drive the animal(s) and one to empty the bucket. Upon reaching the bottom of the incline the animal is either backed up or turned around and walked forward up the incline. In the latter case two animals (or two pairs) are often used. Where two pairs of bullocks are used, three men are needed to operate the arrangement. In such a system one pair walks up the slope while the other is drawing up the bag. As soon as the bag reaches the top of the well it is tipped and emptied into a channel; the bullock driver disengages the toggle that holds the rope to the yoke and the rope is fastened to the yoke of the second pair of bullocks. Although the overall efficiency of this arrangement is poor it has a high degree of reliability (to the farmer this is more important than efficiency).

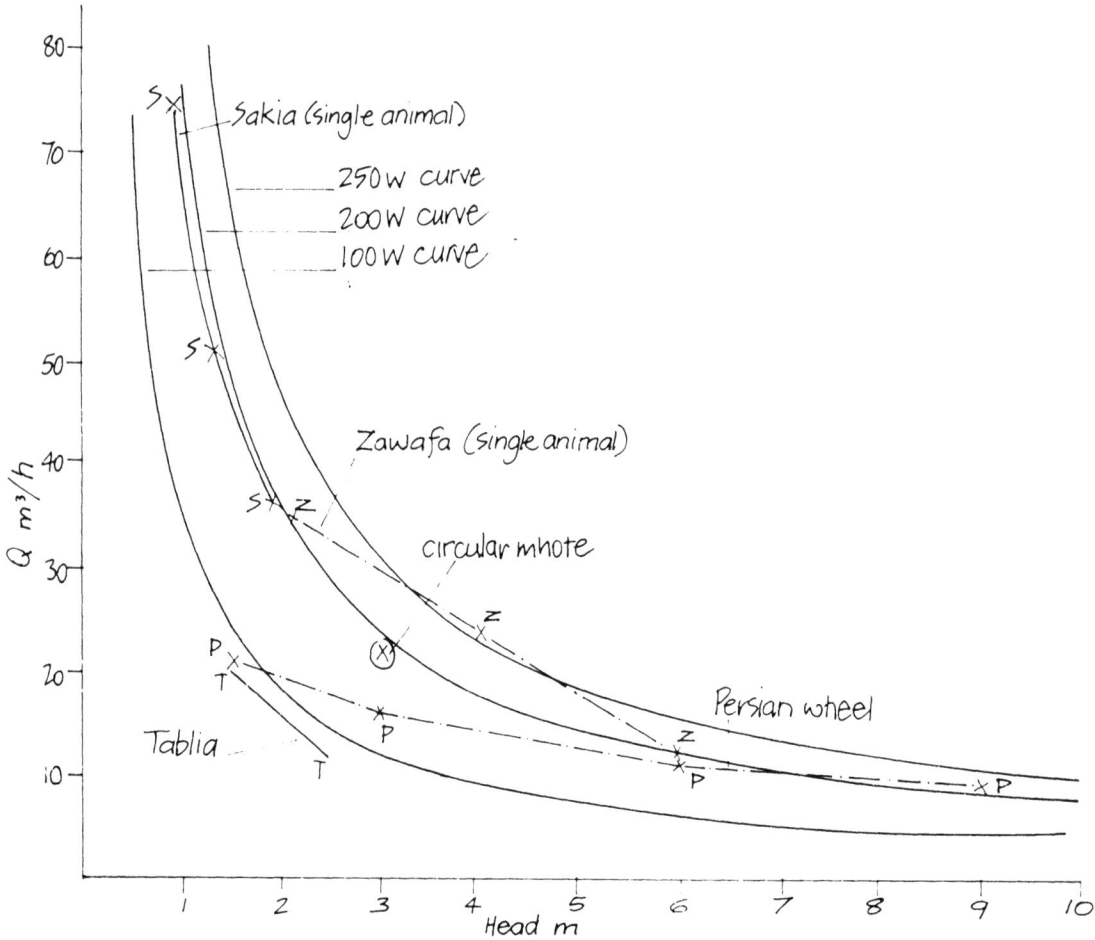

Figure 32. Animal power Q-H characteristics

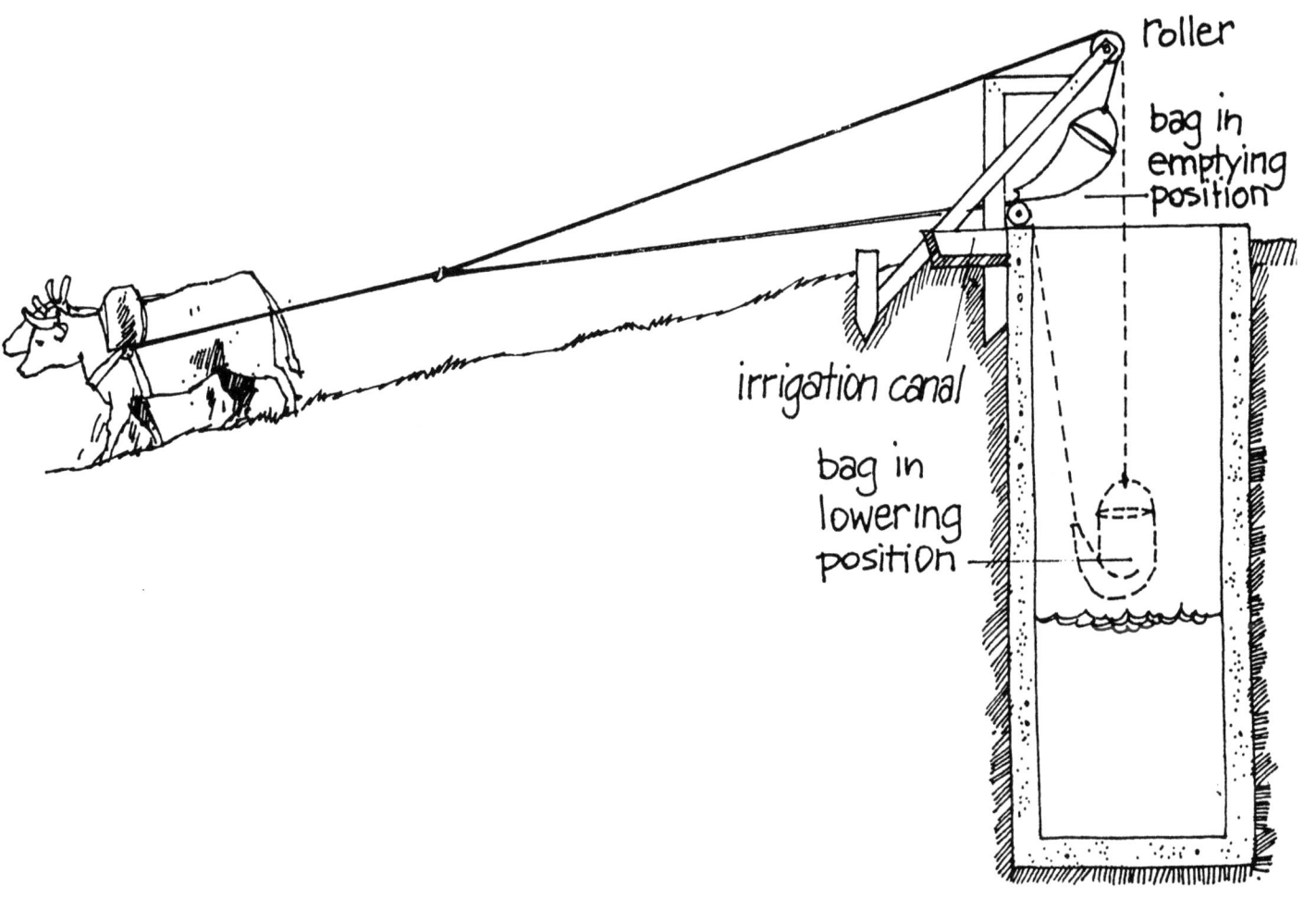

Figure 33. Mohte or rope and bucket (or bag) lift

There are few reliable data on the effectiveness/efficiency of this arrangement. Molenaar (27) quotes a probable average lifting rate of 16 to 17 cubit metres per hour when lifting to a height of about 9m with a two pair animals/three man set-up. He expects about half this discharge rate for a single pair of animals/two man set-up. This is equivalent to a sustained (average) 'water watts' output of 100WW per animal.

4.2.2 The circular mohte

One ingenious variation of the rope-bucket-pulley arrangement is called the circular mohte or 'two-bucket' lift. It is not widely used, mostly in parts of India and Sri Lanka. It uses two buckets - with flap-valves in the bottom - which are alternately raised, emptied, lowered, filled. A chain fixed to each bucket is arranged as shown in Figure 34. The raising and lowering of the buckets is achieved by the chain linkage which passes over an array of pulleys to a horizontal sweep. The sweep is turned in a circular motion by one or a pair of bullocks. The hinged flap-valve in the bottom of the buckets allows them to submerge and fill as they sink into the water. The buckets are tipped automatically at the top of the well and empty into a channel.

This arrangement eliminates the straight line/reversing motion of the simple 'rope and pulley' arrangement and eliminates all (or at least one) of the operators needed for the simple arrangement. Its more complex arrangement (and higher cost) is presumably the reason why its use is restricted to certain areas of India and Sri Lanka.

4.2.3 The Persian wheel

The Persian wheel has long been a popular device throughout the arid zones of the Near East and southern Asia as well as in some Mediterranean countries, for lifting water from wells where the water table is not lower than 9m. The 'wheel' consists of a series of containers - buckets, pots or bags - attached to two parallel loops of ropes or chain which pass over a driving wheel (Figure 35). As the buckets pass over the open-rim wheel, they empty into a trough below. A horizontal shaft extends from the axle of the wheel to a small vertical pinion, which meshes with a large horizontal gear fitted to a vertical shaft; this carries on top a long horizontal beam to which the animal(s) are yoked. It is usually operated by two animals.

A disadvantage of the conventional Persian wheel is the fact that the water is lifted considerably above the surface of the land before it is dropped from the containers into the receiving trough. Note that the effective 'water watts' output falls off significantly with lowering lift partly as a result of this phenomenon. For the performance data indicated on Figure 32 the effective 'water watts' output at

Figure 34. The circular mohte

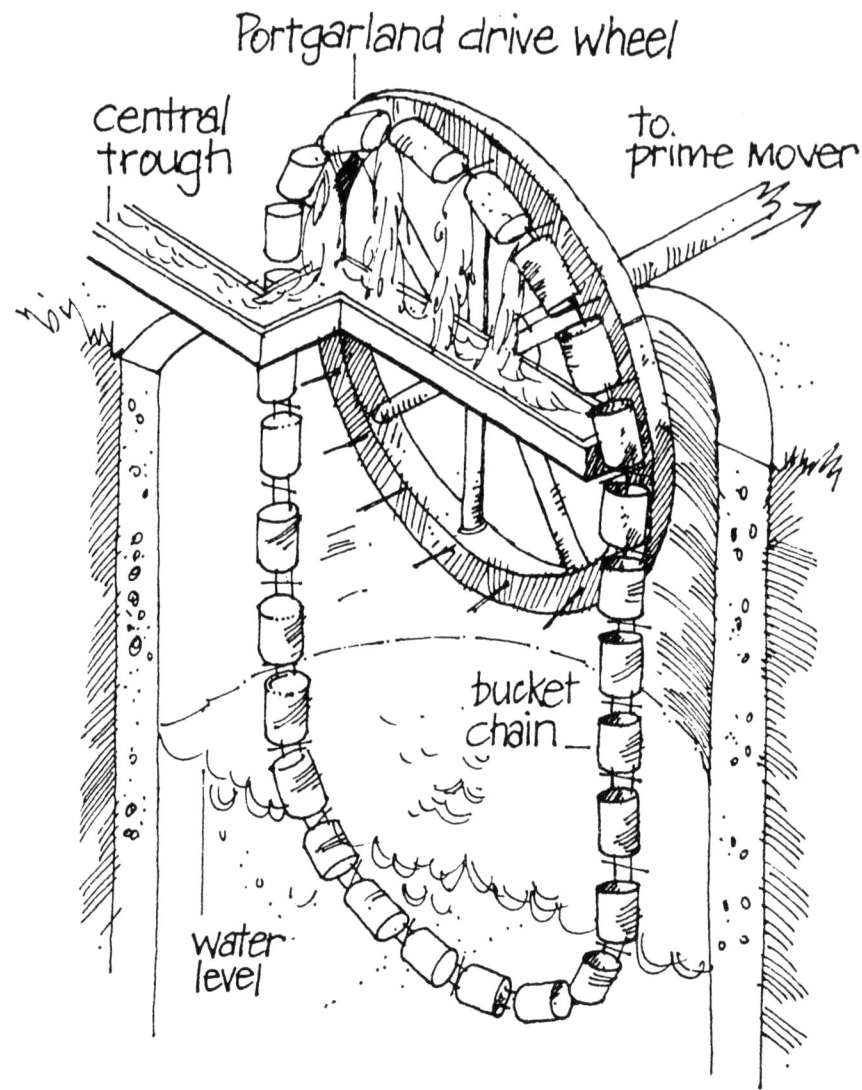

Figure 35. Persian wheel

a lift of 9m is 196W, whereas at 1.5m it is only 82W. This is a serious drawback where the water is close to the surface since this extra lift represents a significant proportion of the total lift. This situation has led to the development of a number of adaptations of the Persian wheel principle, the sakia, the zawafa and the tablia (see following sections).

Persian wheels are usually constructed entirely of wood, except for the containers, but modern versions have begun to use more metal parts. It is reported (48) that Persian wheels with metal gears, bearings, etc., are about 28% more efficient than with wooden components. Schioler (36) reports the efficiency of a wooden Persian wheel with metal bearing surfaces to be about 60%.

4.2.4 Developments based on the Persian wheel

(a) <u>Sakia</u>

A comparatively recent development for lifting water from irrigation channels is a double-sided, all-metal wheel, called a sakia. This has volute-shaped compartments within it. These are open at the periphery of the wheel and curve inwards towards a central opening around the axis (Figure 36). When the wheel rotates, water is scooped up at the outer edge and discharged at the centre into a side trough. Hence no water is being lifted higher than is required. This device has become widely used in Egypt. It has a high discharge rate but its lift is limited to less than half the wheel diameter; it is thus limited to lifts of less than 2m.

The sakia is usually powered by one animal - normally a cow or buffalo, but sometimes a camel or donkey. The animal operating the sakia is changed at the end of a 3-hour period, but each animal may work two shifts a day.

(b) <u>Zawafa</u>

Where a lift of more than 2m is required and the discharge is to be as high as possible with animal operation, a device known as the zawafa (in Egypt) is often installed. Today most zawafas are constructed of metal, although the original versions were wooden. Rather than each bucket delivering water into the same central trough within the driving wheel (as in the sakia), troughs between the spokes of the driving wheel receive the water and direct it to a common trough alongside the wheel. The only limit to the depth from which the zawafa will raise water is the animal power available to drive it. This is usually provided by one or two buffalo.

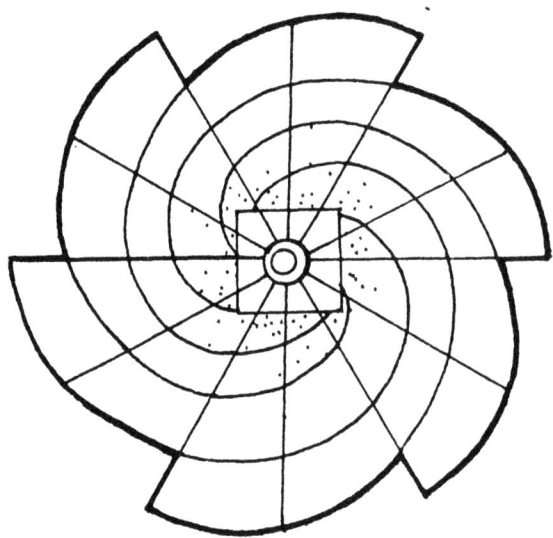

Figure 36. Basis of the sakia water-lifter

(c) <u>Tablia</u>

A device called the tablia is used in those situations where the lift required is too great for the sakia, and the water requirements are not high enough to warrant installation of a zawafa. The device is actually a rim-discharge sakia with the water lifted to a height approximately equal to its diameter less 1.5m.

4.3 Research and development on traditional animal-powered devices

4.3.1 The animal-powered irrigation pump

This pump has been developed at the Punjab Agricultural University, Ludhiana, North India (see their Technical Bulletin - March 1982).

In India the traditional water-lifting devices powered by animals have very low efficiencies. The only exceptions are Persian wheels of improved design (efficiencies of 35 to 45 %). Studies carried out by various institutions have shown that there is a need to develop an efficient animal-driven device.

Various indigenous animal-powered water-lifting devices used in different parts of India are the Persian wheel, the chain pump and the rope and bucket lift. The Persian wheel can be used to lift water up to about 10m and can deliver about $16m^3$/h at a lift of 5m with two bullocks. This is the commonest water-lifter in Northern India. The animal-powered chain-and-washer pump can lift water at the rate of 15 to 20 m^3/h for a lift of 5m with two bullocks. It is used in some parts of Uttar Pradesh state. The rope and bucket lift can lift water at the rate of 10 to $15m^3$/h (lift 5m, two bullocks). It is commonly used in southern India, in the Deccan region and in parts of Rajasthan. To improve upon these traditional animal-powered devices, an animal-powered irrigation pump has been developed at the Punjab Agricultural University, Ludhiana, which delivers $25m^3$/h with a lift of 5m using two bullocks.

The pumping unit consists of two piston-cylinder irrigation pumps. The suction pipes of the pumps are connected by a T-joint. The pumping cylinder is 30cm diameter and the piston stroke is about 10cm (it can be varied). The piston is provided with a rubber sealing washer and there are two non-return valves - a foot-valve at the bottom of the cylinder and one at the piston.

The crank-shaft has a small flywheel at each end. The pump crank is rotated with the pump gear set, having a gear ratio of 3:1. The pump gear set is connected to the bullock gear set through a universal coupling. The flywheel is mounted on the output shaft of the bullock gear set. The bullock gear set consists of a set of gears giving a transmission ratio of 1:36. A wooden beam is connected to the gearbox at one end and the other end is attached to the bullocks with a rope. The overall transmission ratio of the system is 1:12. With a typical animal working pace of 2.5 revolutions per minute the pump speed is 30 revolutions per minute.

The pump can be operated by a pair of bullocks, he-buffaloes or a camel.

Table 6. Animal-owered irrigation pump

Performance (with a pair of bullocks):

Average discharge: $25m^3/h$; suction lift; about 5m
Diameter of walking path: 6.3m
Speed of animals: 2.5 r.p.m.
Draught pull: 50kg
Angle of direction of pull with the wooden beam: $70°$
Input-power: 388W
Output-power: 276W
Overall efficiency: 70%

Economic analysis

Capital cost:
Cost of equipment	Rs*3,500
Life	15 years
Rate of interest	12%
Salvage value	10%
Annual operating hours	2,000
Annual depreciation	Rs 210
Annual average interest	Rs 175
Annual fixed cost	Rs 385
Fixed cost/hour	Rs 0.19

Running cost:

Hiring charge of pair of bullocks	Rs 2.5/hour
Labour charges	Rs 1.5/hour
Maintenance charges	Rs 0.04/hour
Total running cost	Rs 4.04/hour
Total cost of operation	Rs 4.23/hour

*Rs 1 = US$ 0.08

Note that the researchers involved in this work agree that their product is expensive.

4.3.2 Research and development at the Appropriate Technology Development Association, Lucknow, India

A brief report has been received from M.K. Garg of the ATDA at Lucknow regarding research and development on animal-powered pumping there, where on-going work is geared to attaining an increased-discharge animal-powered pump.

Compared to the Persian Wheel ($9 - 11.5m^3/h$ at 6 to 7m head) and the animal-driven chain and washer pump ($13.5m^3/h$ at 6 to 7m), the aim is to obtain a discharge of about $22.5m^3/h$ at 6 to 7m head - the improved performance is necessary in order to meet the increased water requirements of high-yield variety rice. Work has been carried out on a diaphragm pump (the active 'pumping element' is a scooter tyre). It is claimed that a discharge of $20m^3/h$ at 6 to 7m head has been obtained (2-hour stint with a pair of bullocks). Further work is planned with a flexible tube-roller arrangement (details to be obtained).

The aim, as with other developments in this area, should be to obtain an increased discharge without up-marketing the device.

Field study and R and D on animal-powered irrigation devices

There is a dearth of information in the literature on the use of animal power for irrigation. There is a need to seek the views of reliable appropriate technology organizations in developing countries on developments in this area. Although some hardware development has been and is continuing there is a need to obtain a current assessment of this. This assessment must place any proposed developments in this area in the context of present political and socio-economic realities (e.g. the present policies of many states in India is to encourage pump-set tubewell irrigation - with loans given; i.e. the present position does not favour the poorest farmers).

The objective of any project in this area is to achieve a significant increase in pumping capacity without up-marketing the traditional animal-powered devices.

> CAN YOU PROVIDE FURTHER INFORMATION ON SUCCESSFUL DEVELOPMENTS IN ANIMAL-POWERED PUMPING? IF SO PLEASE SEND DETAILS TO THE ADDRESS GIVEN IN APPENDIX 1.

5. DRAUGHT ANIMAL POWER FOR PUMPING: SOCIO-ECONOMIC ASPECTS

5.1 Introduction

This chapter will consider the socio-economic aspects of draught animal power for water-lifting. Unfortunately, as with human-powered devices, there is a shortage of literature specific to water-lifting; most of the writing on animal draught relates to ploughing.

There is a considerable divide between the African and Indian/South East Asian contexts and while there have been numerous oxenization programmes in Africa they have confined the use of draught animals to ploughing.

In considering the use of animal draught for different types of labour tasks such as ploughing, water-lifting and transport, two types of cases are evident. Firstly there is the case of introduction in geographical situations where potential draught animals are not traditionally kept and both the animals and the implements have to be introduced.

Secondly, there is the case of promotion where potential draught animals are traditionally kept but are not used for labouring functions. This must be distinguished from the cases where traditional indigenous water-lifting devices need to be promoted if they are effectively to persist, as in India.

Given that an animal has to be fed incremental fodder for the additional energy expended, the introduction and/or promotion of animal draught for water-lifting would probably only be feasible for irrigation purposes in order to attain increased crop yields. Given that the present subsistence systems tend to use all the residue for cooking, fertilizer and existing fodder the incremental fodder would have to come from increased yields.

5.2 Draught-animal power in Africa

This possible increase in productivity and hence to wealth is a major incentive attracting the small farmer in Africa to the use of animal draught for ploughing. Presumably a similar case can be made for water lifting. Mettrick (26) in his study of oxenization in the Gambia showed how draught oxen achieved only small increases in productivity. This was because agricultural extension was so closely geared to the oxenization programme that women were largely left out: thus one half of the agricultural labour force responsible for a major part of the food supply was ignored.

In many parts of Africa there is a sexual division of labour according to the type of crop, the men being responsible for and receiving the income from cash crops and the women taking care of the subsistence crops for family consumption. Therefore, Mettrick argues, careful consideration must be given to the potential effect on women of the introduction of animal draught. Increased potential for irrigation and concentration on cash crop production may leave women with a greater physical burden in the form of extra labour contributions with the benefits accruing principally to the men.

Murminger (30), who has studied the sociological aspects of using draught animals on African smallholdings, points out that the criteria used by planners as well as farmers for evaluating benefits may differ considerably. Project planners concerned with oxenization have tended to aim for the maximum possible increase in land and crop production. This is limited mainly by the physical capacity of the draught animals but farmers may be satisfied with an increase that is small in comparison with that which is technically possible. The users rate the innovation's capacity to make work easier and to save time higher than full exploitation. Experience on animal traction projects in Africa indicates that actual expansion has been around 20 to 25% of the planners' original projections.

Murminger states, however, that in many African tribal groups the increase in productivity brought about by the introduction of draught power would be an irrelevance. Cattle are envisaged as a medium of exchange to obtain wives, as a means of capital formation and the acquisition of prestige and are commonly imbued with ritual significance. In the African context different basic forms of production usually correspond to ethnic differences and thus to varying languages, myths, rituals and ideologies. The changeover from one type of production to another is therefore not only determined by access to specific resources such as land and livestock but also involves giving up a specific socio-cultural identity in favour of a different one. Technologies are not considered to be neutral with respect to existing cultures and values but are perceived as features of a specific ethnic or socio-cultural identity. This, of course, has a substantial inhibiting effect on the flow of innovations between various ethnic groups. Murminger stresses the importance of investigating these possible constraints during the planning phase of a project aimed at promoting or introducing the use of animal draught.

Oxenization studies draw heavily on the general 'diffusion of innovations' literature. Rogers and Shoemaker (35) devised a set of criteria which, they felt, affected the success rate of the adoption of innovations. In their view the likelihood of an innovation being accepted is much higher if these criteria are fulfilled.

The new practice must be visible to the potential adopters: they must be able to see ox-farming in practice. The introduction of animal draught in the Tanga region of Tanzania (KILIMO programme) failed because it lacked field demonstrations at village level, proper training of animals and proper training of farmers in how to use oxen. The pilot project for oxenization at Mwanyumba village, also in Tanzania (TIRDEP) (40) has selected and defined a target group which will have the benefit of a demonstration farm. Later, farmers practising ox-farming in Mwanyumba will be visible to farmers in other villages.

Potential innovators must be able to test the use of the innovation on part of their fields while continuing with subsistence techniques in other areas. Psychologically this means that they have taken a minimal risk. This choice was not given on the KILIMO programme, the use of animals being exclusively restricted to communal farms. In the TIRDEP project the oxen are to be used on private farms though they are communally owned by the whole village. Several oxenization studies suggest a 'communal approach' where every farmer is able to hire a pair of oxen and can test them out, and the financial risk is shared by the whole community.

An innovation is more likely to be adopted if it is seen by the target group as being relatively simple and uncomplicated. Due to a lack of training of animals and farmers as previously mentioned, the KILIMO programme led the target group to see the technique as being very complicated. To overcome this the TIRDEP project is to take the organizational set-up suggested by the farmers themselves. The smallholder will hire a pair of oxen with a trained labourer. It is hoped that after the implementation of the project and the successful practice of farmers, the innovation will not appear so complex for the farmers of surrounding villages.

The superiority of an innovation over the traditional method must be made clear. In this instance, ox-cultivation has to demonstrate its advantages if farmers compare this technique with hoe-cultivation. It has to show advantages in economic terms of outputs and returns and also prove to reduce the burden of work. The traditional view of cattle as a means of capital formation did not pose a serious danger to the TIRDEP project because only a few Masai hold such views, but communal ownership of oxen would tend to overcome such handicaps as oxen hired out by individual farmers would be regarded as a production factor.

The innovation must be compatible with the present values, traditions and customs. The KILIMO programme suffered because the use of oxen was restricted to communal farms. The TIRDEP approach goes against the traditional pattern of private cattle-keeping but individual farmers can use the oxen on their private fields for their own benefit. The main objection to the communal ownership of cattle is that no-one will feel responsible, or care, for the animals. In the TIRDEP project the labourer, who is employed by the village, is responsible for the oxen put in his charge and must answer to the village government. Unfortunately there appears to be no recent report on the successes and failures of this project, based as it was on the experiences and lessons learnt from a previous oxenization programme.

A GTZ report (13) on the possibilities of introducing draught animals in the North West Province of Cameroon cites a number of factors which are critical to the successful introduction of draught animals. Suitable cattle must be available and there must be no infestation of tsetse fly. There must also be a veterinary service available and enough grazing, both quantitively and qualitively.

This report points to the ways in which the traditional agricultural system is most influenced by the macro-economic environment. Given the crucial role of agricultural policy in developing countries, the technologies that are subsidized and promoted tend to be the most commonly adopted innovations.

An analysis of the Cameroon traditional agricultural system shows that there is subsistence production with extended family structure as the producing and consuming unit, and crops are rotated within a pattern of shifting cultivation. All land is communal and distributed annually according to the needs of the individual and family. As in the Gambia there is a rigid division of labour between the sexes with women cultivating subsistence crops and men growing cash crops.

The GTZ report holds that production techniques and the social conditions of the traditional agricultural structure give little opportunity for the introduction of new farming methods. They therefore feel that a fairly high instability of traditional structure will be a principal social prerequisite for the introduction of draught animals. The report concludes that the introduction of draught animals involves processes of socio-economic change, and successful adoption of such innovations depends on socio-economic and micro-economic factors. Among the socio-economic points to be considered are the instability of the traditional land-use systems and the availability and social security of land. Innovation may be affected by the homogeneity of social stratifications within the target group and possibilities of changing the traditional division of labour. A positive attitude towards the change and cohesive social organization within the group of innovators need to be taken into account.

On a micro-economic level availability of the production factors of land, labour and capital is of particular importance. For the permanent cultivation of land, expansion of farming areas must be possible and credit must be made available to the farmer so that he has capital which will help him to reach a higher production level. The improved income must be enough to allow repayment of credit and the remainder must at least reach the value of the former subsistence production. To reach higher production levels, labour skills must be improved and good training in production techniques is needed.

5.3 Draught animal power in Asia

In India and South East Asia water-lifting devices requiring draught animals are found indigenously and are widespread geographically. Traditional craftsmen construct, maintain and adapt these devices where necessary. Empirical data is primarily available from India and there is evidence that animal-powered water-lifting devices can compete effectively with modern pumpsets under certain conditions. It is therefore fair to deduce that the 'decline' of traditional animal-powered devices is in part due to what may be broadly termed 'political' reasons: the types of technologies that are subsidized and promoted in Rural Development Programmes tend to be more readily adopted even if they are less 'appropriate'. Despite the objectives of India to create rural employment and reduce importation of equipment, the policies implemented by the Indian states make modern devices substantially more attractive. For example, marginal farmers with an upper landholding limit of 2.5 acres receive a 33% subsidy for the cost of installation of a 10 horsepower pump set. A corollary of this is that the bigger farmers benefit considerably more from state policies than do small farmers.

Table 7 shows that since 1974 the number of traditional

devices in Uttar Pradesh, as a percentage of the total number of lift devices in operation has declined.

Persian wheels are still being installed (4838 in 1979-1980) but this is a very small increase (1%) compared to the 16% increase in the installation of pumping sets and private tubewells (175,000 in 1979-1980).

In Uttar Pradesh 1.9 and 2.2 hectares are the minimum size of holdings required for investment in electric and diesel tube-wells. Thus, not only is the small farmer unable to have subsidized access to such devices but he is investing in a considerable surplus capacity which, given his acreage, he cannot use.

Examination of the types of water-lifting devices that have evolved throughout India points to the central influence of the prevailing ecological conditions in the development of these devices. They can be regarded as optimum methods that have been developed and adapted in response to given ecological constraints, hydrological conditions and possibilities for crop exploitation. There are very few studies concerned with the ergonomic aspects of traditional water-lifting devices which, given their evolution and adaptation to local conditions, would probably provide valuable information for ergonomic design. Given the poor track record of hand-pump installations, a strategy to provide an external catalyst as a stimulus to local craftsmen and thus upgrade the relative performance of a traditional device could be socially, culturally and economically the most viable and successful approach, providing that it was felt to be necessary by the local villagers and matched 'external' development objectives.

There is a striking contrast in the evidence as to performance, relative cost-effectiveness and general competitive capabilities of traditional animal-powered water-lifting devices as compared to modern pumpsets. Traditional devices have been criticized because they are only usable under restricted conditions unlike the 'universal' applicability of electric and diesel tubewells. Furthermore the studies of Dhawan and Moorti both put the cost of charsa water considerbly higher than the unit costs of diesel and electric powered tubewells. They estimate that the cost per 1000 cubic metres is: State tubewell Rs.1.4; private tubewell Rs.1.2; Persian wheel Rs.44.6; rope and bucket lifts Rs.93.0. Another study has indicated that the yield from masonry wells is inadequate for the modern demands of irrigation and this has resulted in their decline. This Ghazipur study states that in 1974 Persian wheels and masonry wells occupied a 40% share of the total area irrigated but this dropped to a 9.7% share in 1979.

Table 7. Growth of minor irrigation works in Uttar Pradesh

Achievements up to the Year....	Number of Persian Wheels	Share %	Number of pumping sets	Share %	
1974	4,220,577	43	548,888	57	
1978	455,216	32	948,677	68	(1979)
	460,311	30	1,074,121	70	
1980	465,149	27	1,249,129	73	

Table 8. Comparison of different options at market prices (Rupees)

	Persian Wheel	Electric Tubewell	Diesel Tubewell	Buying	State Tubewell
Capital Cost	2,500	3,500	4,500	N/A	N/A
Lifetime (1)	20	10	10		
Annual Capital Cost (2)	388	544	698	N/A	N/A
Operation Cost	875	522	4,896	N/A	N/A
Maintenance	125	350	450	N/A	N/A
Total annual Cost	1,388	1,416	6,044	7,000	550
Nett annual Cost (3)	1,388	-1,484	4,844	7,000	550

(1) arbitrarily fixed. It should actually be in relation to the number of working hours.
(2) equal instalments, using the diminishing balance formula.
(3) if the excess water is sold.

SOURCE: Vassart (44)

The value attributed to human labour is a critical factor in determining the competitiveness of traditional devices vis-a-vis modern devices. Hurst and Rogers (15) illustrate that with both the Persian wheel and the bullock powered tubewell, on for example a 10m lift (100 days irrigation period and 1 hectare) 500 hours of labour are required whilst a 3 horsepower diesel pumpset requires 83 hours. If family labour is not available to operate the Persian wheel and the bullock-powered tubewell (under the same conditions) they would each require an outlay of Rs.83 (labour valued at Rs.1.0 per hour). As one of the studies explains, the low opportunity cost of surplus family labour and under-utilized excess bullock labour reduces the cost of charsa irrigation below that of alternatives.

Vassart (44) considered the implications of lifting 20,000 cubic metres of water annually which is that necessary to irrigate 1 hectare (this area is thought to be a reasonable one for animal-powered operations) sowed with rice and wheat. He studied five types of water-lifting and his results are shown in Table 8 above.

Vassart argues that the introduction of high yielding varieties of crop and the corresponding increase in water requirements has substantially reduced the viability of Persian wheel and other traditional devices. Further, there is evidence to suggest that in areas where the density of tubewells is high, the yield of masonry wells has fallen. Vassart concludes that for lifts of up to 5 metres and for smallholdings of up to 1 hectare animal powered devices can provide an effective discharge if the annual water requirements are less than 25,000 cubic metres. The situation is highly dependent on the price of oil and the relative merits of traditional devices will improve as oil prices rise. In order to fill the 'gap' between traditional and modern devices Vassart advocates upgrading traditional technologies to increase discharge and reduce costs, develop renewable energies and establish an appropriate organization of co-operative sharing of pumpsets that would fill the void between privately- and state-owned pumpsets.

In the search for the device most appropriate to the needs of the small farmer it is essential not to ignore the choice made by large farmers. The promotion of tubewells in one area will result in a lowering of the water table which will make the traditional devices of the small farmer useless and increase the cost of irrigation water.

In Vassart's view traditional devices are appropriate in situations where bullocks are available, the water table is high and the demand for water is low enough to be met by traditional devices. Only smallholdings of up to 1 hectare will offer scope for using them but since 50% of Indian farmers fall into this catagory then a bullock-powered device can adequately fulfil their needs.

The cost of animal power will be greatly affected by whether the small farmer already owns a bullock or not. If one has to be purchased for pumping, although it can be used for other tasks, some of this capital cost should be included with the pump. In this situation Hurst and Rogers found that the Persian wheel competed effectively with the 3 horsepower diesel pumpset at 5, 10 and 15 metre lifts. However, in the case of complete capitalization then animal draught and the Persian wheel compares unfavourably with the diesel pumpset.

The potential for a communal ownership of bullocks and/or communal investment in a water-lifting device needs to be closely examined. The TIRDEP project in Tanzania (40) opted for communal ownership of oxen and implements which spread the risk among a group of farmers. In many cases the risks may be too high for a small farmer, and the wealthier farmer is in an advantageous position, being able to experiment and having greater access to credit inputs and advice.

Given the massive potential for the exploitation of under-utilized animal power, a strategy to use animal draught would best incorporate both irrigation and ploughing components. Considering the African experience with oxenization it would be more feasible to look at animal draught for water-lifting where animals are already in use for ploughing.

6. IMPACT ASSESSMENT, DIFFUSION OF TECHNOLOGIES, AND MAINTENANCE AND TRAINING PROGRAMMES

6.1 Introduction

The sad record of hand pump performance in developing countries is well documented and is largely attributed to inadequate maintenance provisions.

In Feacham's view (11) the idea that the technology is too complex for rural folk is a myth and broken pumps and inoperative water schemes are found where village skills extend to the repair and maintenance of items such as bicycles and transistor sets. The degree of community involvement and consultation determine what kind of maintenance and repair a pump will receive rather than technological know-how. In order to avoid this situation arising when planners are considering a project they should ask whether the hand pump was the preferred choice of those using the water pump source and also the community at large. They should investigate whether any innovative local developments and knowledge had been ignored when they could have been developed, resulting in lack of motivation or even hostility among villagers.

Another explanation is that a wealth of grass roots technological ability does exist but that local craftsmen deal with qualitatively-different types of technical issues, materials and approaches to design and construction. It is UNICEF's view that, until a deep-well pump can be maintained at the village level, or until local bodies can be upgraded to cope with maintenance issues, only a large-scale government maintenance structure can bring about effective maintenance. This is a pessimistic view and lends weight to the argument that there is a need for a field Research and Development project to investigate the real potential of traditional water-lifting devices, the extent and type of local knowledge and skills and the applicability of these to non-indigenous technical issues. The existence of such a difference of opinion in the literature itself suggests the need for research and development in this area.

6.2 Guidelines for project development

The ITDG Water Panel (17) points out that water-lifting schemes cover a wide range, from a well with a bucket supplying a few isolated huts, to river pumping schemes with supplies piped to several villages. There are obvious difficulties in preparing guidelines to cover such a wide range and this is further complicated by the diversity of socio-economic and climatic factors in developing countries. The Guidelines put most of the blame for failure on poor or non-existent administrative technical support and lack of funds for proper operation. It is also stressed that before selecting an optimum solution, detailed knowledge of the country is required and alternative methods must be evaluated and compared, including all costs and benefits.

The ITDG Guidelines recommend that an appropriate water supply project should be 'more efficient than traditional methods' and should also be simple, robust and reliable, should be supported and accepted by the community and should be organized at the local level with relatively simple training.

Local conditions should also be assessed and regard should be given to available and potential water sources, patterns of water use, health problems (especially those related to water supply), skills of local people, availability of local materials and the institutional infrastructure for management. Sanitation is an important factor which must be taken into consideration. In many situations if more water is made available without adequate sanitation the health of the villagers may suffer. Preference should be given to naturally pure sources which can be protected from pollution.

The Guidelines formulated by the ITDG Water Panel advocate decentralizing control to the village headman or a locally elected authority. They, in turn, delegate to a volunteer caretaker who would have very little assistance but who should always be encouraged. He should be provided with tools and materials and must be fully trained, preferably during the construction and commmissioning stages. The caretaker should be competent to undertake day-to-day routines but regular preventative maintenance should be undertaken by visiting skilled personnel. Manuals should contain operation and maintenance procedures in a simple and concise form. Diagrams and photographs are recommended because they are more likely to be used and consulted.

6.3 Maintenance and training programmes

Accounts of mainstream hand pump projects concentrate in the main on hardware development and/or explaining the different possible models for hand pump maintenance programmes. Maintenance responsibility may fall, at one end of the spectrum, entirely to the central government or may range along a continuum to a shared joint central-local model and on to a village self-reliance model which is unusual. In a system of central maintenance the government commonly assumes all responsibility and meets all costs of well construction, pump installation and the organization of maintenance. Centralization ensures an advantageous exploitation of economies of scale and a minimization of total costs because government purchasing capacities can acquire and distribute spare parts and raw materials, and provide transport and training for personnel. The fact that only a small percentage of hand pumps is in working order at any time suggests that this top-down approach is not very successful.

Towards the self-sufficiency end of the spectrum are the programmes in Kenya and Tanzania where certain villagers, chosen by the community, are given a thorough training in pump maintenance and virtually all responsibility is left in their hands. This person carries a stock of spare parts from the district water office, or if he requires further help he calls on the district engineer's fitters.

Both the WHO International Reference Centre (IRC) (51) and Pacey (32) argue that if a pump could be designed capable of being made by village craftsmen using simple tools and local materials then someone would always be on hand to carry out repairs. This thought is supported by the fact that water-lifting devices of traditional design are built and maintained by village craftsmen, but hand pumps built in villages have not become popular.

The cost factor, in terms of time, personnel input and perceived difficulties and contradictions involved in finding a universal design, has made the approach advocated by the IRC and Pacey generally unattractive to the major donors. The IRC briefly documents cases where traditional designs have been devised that are more suitable to drinking water supply. These, however, have been unsuccessful in intensive community use, lacking durability, and being functionally or structurally inadequate, especially for deep-well application, too expensive and unacceptable to local markets. The IRC states that they are expensive because they tend to be produced in small lots, whereas production needs a critical mass which probably requires an initial subsidy to attain. By all accounts successes with this approach have been few. Pacey cites the example of

Allsebrook's work on the Selekleka pump in Ethiopia. This pump was gradually designed and built in a local context and the villagers were trained to build further pumps. Pacey suggests that aiming for village self-reliance is in itself valuable but in his view it has been approached incorrectly. There is a very plausible view that local skills are attuned to different types of materials, construction methods and technical problems than is western technology. That is to say that the gap is a cultural one: most self-reliance pump designs have more in common with a western do-it-yourself approach than with the style of work done by real village craftsmen.

Large-scale maintenance systems work, in theory, from the premise that there must be no ambiguity in the designation of maintenance responsibilities. Further, they envisage different grades of maintenance and repair tasks, which as they increase in complexity are classified as beyond the scope of the capacities of the village pump caretaker (the bottom end mechanism). Such maintenance organizations are vast bureaucratic structures (frequently of donor agency design and orchestration, such as the UNICEF three-tier system) with numerous levels of authority, involving high inputs of capital and personnel: the recent trend has been to devolve more responsibility onto the beneficiary communities to improve the overall functioning of such systems.

UNICEF (43) accounts of the Tamil Nadu, India three-tier system of maintenance paint a very rosy picture but there is evidence that it is breaking down and changing. At the district level the Tamil Nadu Water Board provides a mobile team for every 1000 pumps although there has been a recommendation that this should be cut to 500 pumps. This mobile team, which includes a mechanic, fitter and helper, is the most expensive component of the system, with large distances involved and high fuel bills which restrict the number of repairs that can be carried out. As a result hand pumps are out of action for long periods and resistance has built up against them in villages. At block level a fitter is provided for every 100 pumps and at village level a caretaker is appointed who is the pivot of the system. His duties are to report promptly to the other levels when repairs are needed, service the hand pump weekly, ensure that it is used properly by villagers, keep the pump area clean and promote good health practices. The village caretaker must have the trust of his fellow villagers and must take his responsibilities seriously. More significantly he must have his own means of support, should be literate to high school level or above and should have some mechanical ability. Likely appointees for this position are nominated by an official, not the community; this does not increase the involvement of the villagers or ensure that the chosen candidate will be acceptable.

However, this system does represent both a change in thinking and a move in the right direction and trends indicate that there is a further devolution of maintenance responsibility as villages take matters more into their own hands. The Tamil Nadu Water Board has increased the drive to develop public awareness of health issues through using traditional organizations such as women's groups. They circulate well-illustrated sets of hand pump maintenance information leaflets. Efforts have been made to raise the prestige of the caretaker in the eyes of the community which increases the likelihood of his advice being heeded and of his pursuing his tasks more diligently.

Despite all the difficulties that have been encountered the concept of large-scale government/agency-sponsored hand pump maintenance systems has not been seriously questioned. However, one Indian state, Rajasthan, has converted to a single-tier system where a selection of villagers are trained as pump mechanics using a government-funded scheme for training rural youth for employment. Responsibility for pump maintenance is handed over to the local council or village leaders.

The fundamental assumption of the UNDP/World Bank Global Project is that hardware development will lead to village maintenance self-sufficiency (VLOM - Village Level Operated and Maintained). As yet there is little information available on the UNDP/World Bank Project but several recent articles in 'World Water' throw light on what can be regarded as the essentials of the Project (41). The concept behind VLOM is that through field-testing, hand pumps will be gradually modified and improved such that eventually it will be technologically possible to operate and maintain them at the village level. The trial projects are in 5 groups and the countries involved were selected on the basis that they have large-scale water supply programmes and could have a 'radiating effect' on their neighbours. These countries are in West Africa, East Africa, South Asia, South East and East Asia and Latin America. There is no great emphasis on software considerations and the trial projects appear to be largely 'laboratory testings in the field'. Only in West Africa is there the intention to study water supply management concepts, software and manpower and training needs. This will be in the Ivory Coast with the additional support of CIDA funds. Unless there is substantial community participation and consultation in the monitoring and adaptation of these pump designs the whole notion of a 'village level operated and maintained' pump may be a nonsense. (See the boxes at the end of this section for more details).

In general, all organizations involved in the sphere of hand pumps in developing countries provide information aimed at the potential pump users and maintenance teams.

This information is usually in written form, but some is presented purely diagrammatically which seems more practical and effective because it overcomes the need for literacy. The maintenance guides present two dimensions to tasks. The first is concerned with the time-scale of tasks, whether it is daily, weekly or monthly, and secondly there is a scale of complexity of the duties. Checklists are provided of possible malfunctions and their symptoms and how these are to be diagnosed and remedied.

Training literature seems to be focused on teaching caretakers about the maintenance of hand pumps. This is largely because it is towards this device that a major part of donor interest and money has been directed. India and Bangladesh have the best documented cases on this subject, stemming from the fact that both governments have invited guidance from international donor agencies. These have chosen to promote hand pumps at the bottom end of the water-lifting spectrum rather than research possibilities for developing traditional devices.

There appears to be only one training programme in hand pump maintenance and repair for women, which was started by UNICEF in Bangladesh. It is difficult to gauge as yet what the programme has achieved or how widely spread it is geographically. Studies tend to suggest that women should be trained, particularly where there are special circumstances such as a high degree of male migration, but there seems to be nothing to show that such schemes have been undertaken.

The oxenization literature pays considerable attention to training schemes, teaching smallholders in the use of animal draught. However these programmes have sought to introduce and promote the use of the plough, and in the African context there was never an attempt to carry out a joint venture in agriculture and irrigation. It seems that whilst a break-in period is necessary to get the farmer and animals used to the plough, harnessing animals to a water-lifting device requires minimal adjustment.

6.4 Impact assessment - the case of the MOSTI

Instances of private and individual hand pump ownership promoted principally for irrigation purposes are numerous across India and Bangladesh. Maintenance is carried out by individual pump owners who receive agency or government subsidies on well construction, hand pump costs and spare parts. While usage on a privately-owned pumpset is less frequent and severe than on a communal one, it is clear that the maintenance record on privately-owned pumpsets is excellent. Rahman and Howes (33) maintain that the MOSTI (Manually-operated Shallow Tubewell for Irrigation) is easily maintained and repaired by Bangladeshi artisans using locally-available materials. This suggests that there is ample local technical initiative and that maintenance problems arise when there is a high degree of government involvement in installation and maintenance of the pumpset, and where there is a collective use of the hand pump.

The apparent popularity of MOSTI among farmers has in turn interested national and international agencies who are particularly attracted by the apparent 'appropriateness' of the innovation to the resources, endowments and technological capabilities of the smaller farmer. In 1975 IRDP and UNICEF established a programme to distribute 63,000 installations over a 3-year period. This was complemented by a further programme organized by USAID to distribute an additional 240,000 pumpsets. The attractions of MOSTI are that it is relatively inexpensive, robust, reliable and easy to maintain and repair by local artisans using materials available in the area. It has a low rate of discharge and a small boro paddy command area (0.5 acres) and is therefore, in principle, best adapted to use on the smallest and most fragmented of landholdings. It is very labour intensive, requiring about one person-year of labour to irrigate one acre of boro rice, and it offers a good rate of return with farmers owning as little as 2/3 acre of land being able to re-coup their initial investment and show a small profit within one 4-month season under normal price conditions.

The inequity effects of the UNICEF, IRDP, USAID and World Bank policies promoting individual MOSTI purchase point to the fact that there is a large section of small holders, and of course the landless, who have failed to benefit at all. The largest number of MOSTI are used by farmers owning 3-5 acres and substantial numbers are owned by bigger farmers. The World Bank attempted to stipulate that MOSTI should not be supplied to farmers owning more than 0.75 acres, but this had to be referred back on the grounds that the credit repayment scheme could never be successful because the smaller farmers could not generate enough surplus to meet the repayments, and were less able to take risks in the first place. It is these groups, who have not been able to benefit from MOSTI schemes, who could gain most from efforts to examine the viability of promoting traditional water-lifting devices and systems of appropriate collective ownership and use.

In Rahman and Howes' view (33) farmers will only switch from boro to other crops, which need less water and have a better nutritional value, if there are clear marketing opportunities. Farmers will only have labour for pumping if its opportunity cost is relatively low so the innovation holds out few employment prospects for the landless labourer. Women will not participate in pumping even if there is a domestic labour shortage.

Rahman and Howes' assessment suggests that MOSTI has an important contribution to make given that it is relatively efficient in its use of scarce resources and also more suitable for small farmers. However, MOSTI in its present form does not provide the answer for most of the small and most impoverished farmers in Bangladesh. The interests of this group will only be served by the selection or generation of even cheaper and possibly more labour-intensive techniques.

The development of the bamboo tubewell was an attempt to provide cheap irrigation. It was invented in Bihar, North Eastern India around 1969 following a government programme encouraging irrigation. There wells cost Rs.4000 and a further Rs.4000 for the pump, and credit was only available to larger farmers. Hence farmers experimented with cheaper materials, using bamboo as a lining and bamboo and coir from coconuts to make the strainer. The only non-local materials were the coir and some strips of galvanized iron to hold the casing in place. The wells went down 70 to 80 feet and boring, which was done by the sludge method, took 20 to 25 days. All the necessary skills were available locally and the cost was Rs.100 to Rs.160. A further innovation, of bolting the diesel pumpset to ox carts, enabled the pump to service several wells and the owner could increase the rate of return on his investment by hiring out his equipment. After some initial disapproval, government agencies started to offer credit for pumps used with bamboo tubewells and there was a rapid rise in their numbers: 154 in 1968/9; 1,200 in 1970/71; and 19,000 in 1972/3.

However, despite its relative cheapness and apparent success very few small farmers benefited. Only a few wells were installed by farmers with 2.5 - 5 acres, who are by no means the poorest, and only 10% of the wells were installed by farmers with less than 10 acres of land. The poorer farmers had very small and distant plots which were difficult to irrigate even with a mobile pump set.

This is an example of a development, using local materials and skill, which caught on very rapidly. Nevertheless it once again illustrates how the poor may be debarred from benefiting from new technology by such factors as the distribution and size of land plots and the lack of loan facilities.

Appendix 1

Constructive comments on this report - including further information, details of developments in water-lifting, field projects, etc. are welcomed by the Intermediate Technology Development Group. As mentioned in the introduction this report was intended to be an 'ideas generator'. It is hoped it will be a working document and will stimulate comment, discussion and further field developments.

The address to write to is:

Intermediate Technology Development Group Ltd.
9 King Street,
London   WC2E 9HW.
U.K.

Appendix 2

References

1. AGARWAL A., 'Supplying Water: Maintenance Problems and Community Participation', Development Digest, October 1981.

2. Battelle Report to AID 'The Development of a Water Pump for Developing Countries', Battelle Memorial Institute, 1962.

3. BURGEAP, 'The Construction of Wells in Tropical Africa and the Human Investment: Niger, Upper Volta and Tchad', Series Techniques Rurales en Afrique 1964-1973.

4. BURTON J.D., 'Joggle Pumps using the Induced Flow Principle' Proceedings of the Sixth Conference on Fluid Machinery Vol.1, Hungary Academy of Sciences, Budapest, 1979.

5. CHAMBERS, R., 'Irrigation and Agricultural Development in Asia',

6. CHAUHAN, S.K., et al., 'Who puts the water in the taps?' An Earthscan Paperback, International Institute of Environment and Development, 1983.

7. COHEN and UPHOFF, 'Participation's Place in Rural Development: Seeking Clarity through Specificity', World Development 1980.

8. Consumer Association 'Hand/Foot Operated Water Pumps for use in Developing Countries', Testing Research Report (Z9923), Part 1, Part 11, Sept. 1979; Final Summary Report, Oct. 1980.

9. COWARD, E.W. Jr., (Ed)., 'Irrigation and Agricultural Development in Asia; Perspectives from the Social Sciences', Cornell University Press, 1980.

10. DUBE, S.C., 'Cultural Factors in Rural Community Development', Journal of Asian Studies 16, 1956.

11. FEACHAM, R.G. 'Community Participation in Appropriate Water Supply and Sanitation Technologies: the Mythology for the Decade', in 'More Technologies for Rural Health. Proceedings of the Royal Society', London, 1980.

12. FAO/UNDP, Proceedings of the China Workshop on Water-Lifting Devices and Water Management, China, Nov. 1981.

13. German Agency for Technical Co-operation, 'Possibilities of the Introduction of Draught Animals in the North-West Province of Cameroon', GTZ Report.

14. HOLMBERG, A.R., 'The Wells that Failed: an attempt to establish a stable water supply in Viru Valley, Peru', Human Problems in Technological Change, a Casebook, E.H. Spicer, ed. 1952.

15. HURST, C. and ROGERS, P., 'The Bullock Powered Tubewell: An Economic Analysis', 1981.

16. Institute of Development Studies, Planners Preferences and Local Innovation in Tubewell Irrigation in N.E. India', Discussion Paper No. 70, 1974.

17. ITDG Water Panel, 'Guidelines on the Planning and Management of Rural Water Supplies in Developing Countries', Appropriate Technology, Vol. 7, No.3, December 1980.

18. International Rice Research Institute, 'IRRI Bellows Pump: a simple way to pump water', Undated leaflet.

19. ISELY, R., 'Facilitation of Community Organisation', WASH Technical Report No. 7, 1981.

20. JOURNEY, W.K., 'INT/81/026, Hand Pump Testing and Development, Progress Report for South Asia Region', Dec. 1983.

21. KHAN, H.R. (Country Statement, p.227) FAO/DANIDA Workshop on Water-Lifting Devices in Asia and the Near East, Dec. 1979.

22. KHARE, R.S., 'A Study of Social Resistance to Sanitation Programmes in Rural India', The Eastern Anthropologist, 1964.

23. KHEPAR, S.D., KAUSHAL, M.P. and MANGAL SINGH, 'The Animal-Drawn Lift-Irrigation Pump'. Technical Bulletin, Department of Soil and Water Engineering, Punjab Agricultural University, Ludhiana, India, 1982.

24. Kingdom of Thailand - Ministry of Interior, 'Acceleratetd Rural Development Office Handpump Improvement Project'.

25. MATANGO, R.R., 'Maji na Maendeleo Vijijini: The experience with rural self-help water scheme in Lushoto District', Water Supply in East Africa, April 1971 Proceedings of the Conference on Rural Water Supply, G. Tschannerl ed., University of Dar es Salaam, Bureau of Resource Assessment and Land Use Planning, Research Paper 20.

26. METTRICK, 'Oxenization of the Gambia', 1977.

27. MOLENAAR, A., 1956, p.21. 'Water Lifting Devices for Irrigation', FAO Agricultural Development Paper No.60.

28. MORGAN, P.R., 'A Simple Low-Cost Water Pump for Shallow Wells', Appropriate Technology, Vol 8, No.3, Dec. 1981.

29. MOYES, A., 'The Poor Man's Wisdom', Oxfam 1979.

30. MURMINGER, P., 'Animal Traction in Africa. Sociological Aspects of the Use of Draught Animals on African Smallholdings', GTZ, 1982.

31. NAZRUL ISLAM, S.M., 'Energy Recovery in Man-Powered Pumps', Vol.14 No.1, Agricultural Mechanisation in Asia, Africa and Latin America, 1983.

32. PACEY, A., 'Handpump Maintenance and the Objectives of Community Well Projects', Oxfam Socially Appropriate Technology Manual No.1, 1976.

33. RAHMAN, M., HOWES, N., 'A Preliminary Assessment of the Determinants, Consequences and Policy Implications of Handpump Irrigation in Bangladesh'.

34. RAVEENDRAMPILLAI, V., 'Archimedean Screw Pumps'. Project Report, Department of Civil Engineering, Middlesex Polytechnic, 1980.

35. ROGERS, E.M., SHOEMAKER, F.,'Communication of Innovations: a cross-cultural approach'. 2nd Edition. New York, Free Press, 1971.

36. SCHIOLER, T., 'General Background Paper on Water-Lifting Devices for Irrigation', Proceedings of FAO/DANIDA Workshop on Water-Lifting Devices in Asia and the Near East, Bangkok, 1979.

37. SPARE D., 'The Rower Pump: A Summary of its Development 1978-1982', Mennonite Central Committee, Box 785 Dhaka-2, Bangladesh, 1982.

38. Tamil Nadu Water Board, 'Guide to the Selection and Training of Village Handpump Caretakers'.

39. THANH, N.C., PESCOD, M.B. and VENKITACHALAM, T.H., 'Design of Simple and Inexpensive Pumps for Village Water Supply Systems', Asian Institute of Technology, Environmental Engineering Division, Bangkok. Final Report No.67, Jan.1977.

40. TIRDEP, 'Pilot Project for Oxenization at Mwanyumba Village. Results and Recommendations of a Sociological Survey' Tanzania.

41. UNDP/World Bank Global Project, World Water, 1983.

42. UNDP, 'Low Cost Groundwater Development', Manual for African Regional Seminar, Lilongwe, Malawi, 1982.

43. UNICEF, 'Caretakers for Handpumps in Tamil Nadu, India', Assignment Children.

44. VASSART, M., 'Technologies for Lifting Irrigation Water' ILO Special Paper, 1981.

45. WATT, S., 'Chinese Chain and Washer Pumps', Intermediate Technology Publications Ltd., 1976.

46. WATT, S., 'The Mechanical Failure of Village Water Well Pumps in Rural Areas', Appropriate Technology, Vol 4, No 3.

47. WHYTE, A., BURTON, I., 'Water Supply and Community Choice' in Water, Wastes and Health in Hot Climates. Feacham, McGarry and Mara (ed.) John Wiley and Sons, 1978.

48. WOOD, A.D., RUFF, J.F., RICHARDSON, E.V. 'Pumps and Water Lifters for Rural Development', Colorado State University, 1977.

49. World Bank, 'Rural Water Supply Handpumps Project', Report No 1 March 1982.

50. World Bank, 'Rural Water Supply Handpumps Project', Technical Paper No.6 March 1983.

51. World Bank Technical Paper No.19, 'Rural Water Supply, Handpumps Project Final Technical Report', June 1984.

52. WHO International Reference Centre for Community Water Supply, and Sanitation 'Strategies and Management Structures for Maintenance of Handpumps in Rural Water Supply Programmes in Developing Countries', Draft, The Hague.

53. WHO International Reference Centre for Community Water Supply and Sanitation 'Handpumps for use in Drinking Water Supplies in Developing Countries' Technical Paper No.10, The Hague, July 1978.

www.ingramcontent.com/pod-product-compliance
Ingram Content Group UK Ltd.
Pitfield, Milton Keynes, MK11 3LW, UK
UKHW050226150426
5217IPUK00023B/1671